GLOBAL WARMING
FALSE ALARM

THE BAD SCIENCE BEHIND THE UNITED NATIONS' ASSERTION THAT MAN-MADE CO_2 CAUSES GLOBAL WARMING

GLOBAL WARMING
FALSE ALARM

THE BAD SCIENCE BEHIND THE UNITED NATIONS' ASSERTION THAT MAN-MADE CO_2 CAUSES GLOBAL WARMING

RALPH B. ALEXANDER

CANTERBURY PUBLISHING

Library of Congress Control Number: 2009905563

Global Warming False Alarm: The Bad Science Behind the United Nations'
Assertion that Man-made CO_2 Causes Global Warming
Alexander, Ralph B.
Includes bibliographic references and index.
ISBN-13: 978-0-9840989-0-3 (pbk. : alk. paper)
ISBN-10: 0-9840989-0-9 (pbk. : alk. paper)

Published by Canterbury Publishing
P.O. Box 1731, Royal Oak
Michigan 48068-1731
SAN 858-4133

Book design by Karl Rouwhorst

Printed in the United States of America
The paper used in this publication meets the minimum requirements of the
Permanence of Paper standard ANSI/NISO Z39.48-1992 (R2002).
10 9 8 7 6 5 4 3 2 1

Contents

Preface

I wrote this book because I'm a scientist. Because I'm offended that science is being perverted in the name of global warming – today's environmental cause célèbre. Because the world seems to have lost its collective mind and substituted political belief for the spirit of scientific inquiry.

Science is not a political belief system. Yes, scientists are human and have their biases, but the keystone of science is rational investigation. There's nothing very rational nor investigative about much of the conventional wisdom on global warming, which is characterized more by a near religious zeal than by thoughtful evaluation of the evidence.

It's the abuse of science by global warming alarmists that turned me into a skeptic about CO_2. Until fairly recently, I was a fence sitter and willing to accept the possibility that recent climate change is the result of human activity, though I also had my doubts.

What first changed my view was a college course on physical science that I taught a few years ago and that included a segment on global warming. Despite the fact that the course focused on the scientific method, and on not accepting hypotheses without adequate testing or evidence, the textbook simply presented the alarmist line on man-made global warming without question.

To me, that made a mockery of the history of science presented in the course, which featured several examples of how mainstream scientific thinking has sometimes been wrong in the past. At the very least, I felt that the other side of the global warming debate should have been discussed as well.

The experience induced me to take a second look at global warming. As I delved more deeply into the background material, I found myself steadily moving over to the skeptical camp and becoming more and more annoyed at the strident tone of most alarmist declarations – especially the assertion that "the debate is over". Such an assertion would have appalled the famous 19th-century British biologist Thomas Huxley, often regarded as one of the founding fathers of modern scientific thought, who once said: "Skepticism is the highest of duties; blind faith the one unpardonable sin".

This book takes the skeptics' viewpoint on global warming, but with emphasis on the underlying science rather than the politics. My intention is as much to resurrect the tarnished reputation of scientific endeavor as it is to convince you that CO_2 has little to do with global warming.

Why have I taken it on myself to defend the honor of science?

It's because I feel so strongly about my chosen profession. I've been interested in science since my childhood, when science was held in particularly high regard following the development in only a few years of nuclear power, the transistor (the basis of today's computer chip), and rockets capable of sending man to the moon – along with advances in the biological sciences such as figuring out the structure of DNA.

Some of the gloss of those heady times later wore off, and there was an understandable backlash when it was realized that science didn't have all the answers to our problems. Unfortunately, this reaction helped fuel the rise of "junk science", based on ignorance and fear instead of the traditional scientific method. And all that coincided with the start of the global warming debate about 20 years ago, so perhaps we shouldn't be surprised that the debate has generated more heat than light.

Much of the book is critical of the UN's Intergovernmental Panel on Climate Change (IPCC). That's because the current belief that global warming is caused by man-made CO_2 stems largely from the IPCC's climate assessment reports, as interpreted by the mass media and swallowed by the public at large.

More than anything else, it is the IPCC reports that have convinced me alarmists are wrong about CO_2.

I'm not saying that the lengthy, detailed reports are full of bad science. They do contain accurate and useful information in places, supplied by well-meaning climate researchers who are genuinely trying to piece together a coherent picture of the Earth's climate system. But these scientists appear to be in the minority within the IPCC, which is dominated by other scientists and bureaucrats who manipulate the data and the reports for their own ends.

All of this is a sad commentary on the state of science today. It has taken more than two millennia to develop and refine the modern scientific method to the point where we've been able to make major technological advances in a relatively short time. But if we continue to debase science as the IPCC and other political bodies do, our wonderful scientific heritage will be lost.

The book is written for both the layman and the scientist, at a level that anyone with a high-school education including some basic science should be able to understand. However, even those who can't comprehend the science in detail should be able to follow the general line of argument.

To convey the essence of global warming science in a readable and informative manner, I've kept technical material and scientific jargon to a minimum in the main text. Scientists, and those nonscientists seeking more detail, will have to dig a little further by consulting the many endnotes (and the appendices) at the back of the book.

There are several people I want to thank. These include my brother Patrick, who first got me seriously interested in global warming and encouraged me to write this book; my colleagues Keith and Hillary Legg, for many discussions on the subject and useful suggestions; Jim Peden, who convinced me to actually start writing; Howard Hayden and Roy Spencer, for answering my questions; and my wife Claudia, for her unwavering support and encouragement as I took on what is still an unpopular cause.

Chapter 1: Global Warming Deceit

If you believe what you read and hear in the media, we've only got ourselves to blame for global warming – it comes from the carbon dioxide (CO_2) that we're pumping endlessly into the atmosphere by burning fossil fuels. Those who don't accept this mainstream view are dismissed by climate change alarmists as ill-informed, or even labeled as environmental criminals.

But is the conventional wisdom correct?

To global warming believers, the climate bible is a series of reports issued by a United Nations body, the Intergovernmental Panel on Climate Change (IPCC). Based on the collective opinion of several hundred climate scientists, the IPCC reports are the source of the widely held belief, promulgated by Al Gore and other alarmists, that higher temperatures are the result of human activity.

The IPCC and global warming alarmists spare no effort in telling us to fix the problem by curtailing or even ending emissions of CO_2. Taking action on CO_2 is essential, the believers shout, because temperatures will increase further, with drastic effects on our planet and our way of life unless we do something right now. If we don't, goes the man-made CO_2 story, the Earth is in for more weather extremes, big shifts in rainfall patterns, and thawing of the polar ice caps that will cause devastating rises in sea level.

This is all nonsense. Global warming may be real, but there's hardly a shred of good scientific evidence that it has very much to do with the amount of CO_2 we're producing, or even that temperatures have risen as much as the IPCC says.

You wouldn't know that from following the news. The alarmists have spun such a web of deception that any science contrary to the view of human-induced global warming is either ignored, played down, or deliberately distorted. News releases and scientific papers that don't adhere to the IPCC "party line" on CO_2 are immediately sidelined by a barrage of attacks, sometimes vicious and personal.

ALARMISTS AND SKEPTICS

Global warming skeptics have very different ideas about the origins of increasing temperatures. Unlike alarmists, skeptics believe that humans have little, if anything, to do with global warming and dispute the notion that our CO_2 emissions have any significant effect on climate. They question the basic science behind the whole case for a human influence on the climate system.

To skeptics, global warming is almost entirely a result of natural causes. Therefore, there's no point in passing legislation on CO_2 emissions arising from the human presence on Earth – at least, no point in doing it to combat global warming.

Both groups defend their views vehemently, although the debate, if it can be called a debate, is mostly conducted out of public sight in the silent world of Internet blogs. The public, visible picture is quite different. Almost exclusively, the mainstream media present the alarmist viewpoint alone and do so as if man-made global warming were an established fact, a belief that no longer needs to be questioned or debated. This standpoint has even been adopted by many professional scientific societies.

The skeptics, frustrated, struggle to be heard. This is not only because their opinions and publications on the subject are not taken seriously, but also because alarmists are very vocal in putting their side of the case.

Although there are signs that the tide is turning, skeptics are still denigrated and even publicly vilified by alarmists. Until recently, the alarmists delighted in bad-mouthing anyone who didn't subscribe to their convictions by calling them "deniers" – an attempt to link global warming skeptics with the immorality of Holocaust deniers. Although this tag failed to stick for long, it illustrates the emotional intensity of the debate.[1]

Dr. Rajendra Pachauri, who is the IPCC chairman, has ridiculed those who question the man-made global warming orthodoxy by comparing skeptics to members of the Flat Earth Society, which he said[2] probably has about a dozen members today. Former U.S. Vice President Al Gore has made similar statements, claiming there is a misconception that the scientific community disagrees over global warming, and insisting that a scientific consensus exists.[3]

It is Al Gore and the IPCC who harbor a misconception, as the number of skeptics is far from small. Exactly how many scientists are on the alarmist or skeptical side of the global warming debate is uncertain for various reasons, but more and more skeptics are now coming out of the scientific closet.

Scientists

In December 2007, U.S. Senator James Inhofe held a Senate hearing on global warming and the media, in an effort to balance the media's one-sided stance in presenting the science of climate change. The hearing's official report[4] included a list of about 400 scientists from over 20 countries who have voiced significant objections to the so-called consensus on human-caused global warming. The list of dissenting scientists is constantly updated, and had grown to more than 700 by March 2009.[5]

Critics have pointed out that not all those on the list are scientists. It includes, for example, some 25 economists, although economists are also among the authors and reviewers of the IPCC's climate reports. Yet even if the nonscientists are excluded, that leaves well over 600 scientists – including several associated with the IPCC itself – who dispute the climate claims made by the IPCC. And that may be only the tip of the iceberg since, as the original Senate report says, "Many of the scientists featured in this report consistently stated that numerous colleagues shared their views, but they will not speak out publicly for fear of retribution."[4]

The use of intimidation to silence skeptics on man-made global warming in the scientific community has been observed before, notably by prominent climate scientist Richard Lindzen, who is Alfred P. Sloan Professor of Atmospheric Science at MIT:

> But there is a more sinister side to this feeding frenzy. Scientists who dissent from the alarmism have seen their grant funds disappear, their work derided, and themselves libeled as industry stooges, scientific hacks or worse. Consequently, lies about climate change gain credence even when they fly in the face of the science that supposedly is their basis.[6]

As well as skeptical scientists who keep quiet out of fear, there are others who feel no particular need to make their views known in public. I was in that category myself before embarking on this book. What all of this means is that any count of scientific skeptics on global warming is bound to be an underestimate.

An opinion poll of over 500 *climate scientists* worldwide on the origins of global warming was carried out by Dennis Bray and Hans von Storch, first in 1996 and again in 2003.[7] The 2003 survey, conducted online, found that almost twice as many of the climate scientists polled believe that climate change in general is due to human activity as those who do not.

Because a specific question about global warming was not asked in either poll, it is hard to draw definite conclusions from the results. Nevertheless, it is clear that a substantial number of climate scientists, even if not a majority, are global warming skeptics – about 35% in this study.

A more recent attempt to take the pulse of the global warming debate was a survey of 3,146 earth scientists in 2008.[8] This survey did ask respondents specifically if they thought human activity was contributing significantly to changing temperatures, a question to which 82% of those surveyed answered yes. This suggests that only 18% of earth scientists are global warming skeptics, although the same poll pegged earth scientist skeptics who regard themselves as climatologists at an even lower level, which appears to conflict with the Bray and von Storch survey.

An ongoing project designed to attract skeptics on global warming is the Oregon "Petition Project".[9] It was originally organized in 1998 by the Oregon Institute of Science and Medicine, in order to counter the then-widespread claims that an overwhelming number of scientists agreed with the hypothesis of human-induced global warming, and that further examination of the science was unnecessary. The claim about the "overwhelming number of scientists" supporting their cause is still prevalent among alarmists today.

More than 31,000 U.S. scientists have signed the Oregon petition. However, it refers to "catastrophic heating" of the Earth's atmosphere and "disruption" of the earth's climate, rather than global warming as such. So, as in the surveys of climate scientists that did not ask explicitly about global warming either, the exact number of global warming skeptics among the Oregon petition signatories is unclear. But what is evident again is that the number is large.

A widely quoted essay that appears to support the IPCC-Al Gore claim that global warming skeptics are small in number was published in *Science* by Naomi Oreskes, a history professor at the University of California-San Diego, in 2004. Oreskes analyzed 928 abstracts of papers related to "global climate change" published in refereed scientific journals between 1993 and 2003, and listed in a scientific database.[10] She claimed to have found that none of the 928 abstracts disagreed with the assumed consensus position among climate scientists on man-made global warming.

Unfortunately for the alarmists, Oreskes' finding on the apparent absence of skeptical views is flawed. A subsequent analysis of the same set of abstracts[11] came to the quite different conclusion that a majority made no mention of man-made global warming at all. In any case, even if none of the other papers refuted the assumed consensus position on human-induced warming, it's likely that none of them said the warming couldn't be due to natural causes either.

Nonscientists

In the public at large, the percentage of skeptics is generally higher than among scientists, according to recent polls – around 50% both globally and in the U.S. (Table 1.1).

An average of 54% of respondents to a Gallup Poll conducted across the world in 2007 and 2008 think that "rising temperatures ... are a result of human activities", with the alarmist percentage in the U.S. (49%) being slightly below the worldwide average. A Pew Research Center survey in January 2007 came up with a similar result for the U.S., just 47% of the public believing that the Earth is getting warmer due to human activity such as the combustion of fossil fuels.[13] While these numbers are large, they also indicate an almost equal number of global warming skeptics, both in the U.S. and worldwide.

Table 1.1: The Global Warming Debate

	Alarmists	Skeptics
Climate scientists, world[7]	65%	35%
Earth scientists, U.S. and Canada[8]	82%	18%
General public, world[12]	54%	46%
General public, U.S.[12]	49%	51%

All these polls therefore give the lie to the IPCC chairman's declaration that global warming skeptics barely exist. Although perhaps less numerous than alarmists, at least among scientists, skeptics abound. My hope is that this book will create more skeptics yet.

CORRUPTED SCIENCE

As a scientist, what I personally find most troubling about the global warming debate is the gross misuse of science by those on the alarmist side. The worst public offender is the IPCC, which long ago became the accepted authority for those convinced that global warming is human-induced.

In this book, I will expose the most flagrant abuses of normal scientific practice by the IPCC, and draw attention to the questionable assumptions and interpretations of data that the IPCC has used to formulate its position on global warming.

This will involve looking at both the science – which we'll do at a fairly basic level – and the methodology used to interpret the available climate data. Good science is based on what is called the scientific method, a set of procedures established over many centuries since the time of the ancient Greeks. In later chapters, we will see where the IPCC has not only gone wrong in its handling of climate data, but has also departed repeatedly from sound scientific methodology – to the point of corruption in many instances.

But first, let's review what the IPCC is and what it has to say about global warming.

THE INTERGOVERNMENTAL PANEL
ON CLIMATE CHANGE (IPCC)

The IPCC has always believed that global warming is man-made, ever since it was established 20 years ago. The organization was founded jointly by the UN Environment Programme and by the World Meteorological Organization, a group that works to standardize weather-related observations. The IPCC's stated purpose was to assess "the scientific, technical and socioeconomic information relevant for understanding the risk of human-induced climate change".

But despite the implication in this statement that climate change (read: global warming) caused by human activity was still in the future, it's no secret that the UN and the IPCC believed it was already happening. Though cloaked in bureaucratic language, this presumption can be clearly seen in the terms of the mandate that included, in addition to wording on filling gaps in our climate knowledge, statements such as:

> Review of current and planned national/international policies related to the greenhouse gas issue; Scientific and environmental assessments of all aspects of the greenhouse gas issue and the transfer of these assessments and other relevant information to governments and intergovernmental organisations to be taken into account in their policies ...[14]

In other words, the whole IPCC effort has always been biased toward the assertion that humans alone have caused the warming we now measure across the globe. This bias is a central part of all the panel's publications and reports. By the IPCC's own admission,[15] even the first report in 1990 "... made a persuasive, but not quantitative, case for anthropogenic [human-caused] interference with the climate system."

Subsequent IPCC reports have reinforced the original conclusion, adding more and broader arguments for the existence of man-made warming, and putting numbers to this and other assertions (Table 1.2). Each successive report has sounded more and more confident that global warming is largely our own fault. By the time the Fourth Assessment Report was issued in 2007, the IPCC claimed to be up to 90% certain of its conviction that rising temperatures are caused by human activity.

All of that would be unimportant were the IPCC not so powerful: numerous governments around the world, not to mention environmental groups and the general public, regard its word as climate gospel – to be taken as the absolute truth, without question. Its reports on global warming are far more widely read and quoted than most speeches by world leaders on any topic at all. The IPCC itself rightfully claims that its reports immediately become "standard works of reference". This is an enviable position that most interest groups and professional societies can only dream of.

If the IPCC were simply an organization of climate scientists without any agenda other than to review and understand the many factors affecting the Earth's climate, its pronouncements might be more believable. But the reality is that the IPCC also consists of a lot of bureaucrats and government representa-

TABLE 1.2: PRINCIPAL IPCC ASSERTIONS[16]

Assertion	Confidence level
1. Man-made CO_2 and other greenhouse gases[17] in the atmosphere have increased significantly since 1750.	100%
2. Most of the global warming in the last 50 years has been caused by these gases.	At least 90%
3. Temperatures in the Northern Hemisphere since 1960 have been the highest of any 50-year period in the last 1,300 years.	At least 66%
4. Even greater warming will occur this century if we continue to emit CO_2 and other greenhouse gases.	At least 90%

WHAT'S WRONG WITH THIS PICTURE?

- Assertions 2, 3 and 4 are debatable (though assertion 1 is not in dispute) and may not be true at all.

- Confidence levels as high as 90% are totally unjustified, because these three conclusions are based solely on computer models that are only crude approximations to the Earth's climate.

  Such high confidence levels have led to the false, unsupportable beliefs that there is "scientific consensus" on human-induced global warming and that "the science is settled".

  The phrase "global warming gases" has become part of everyday usage, despite the lack of any proven connection between greenhouse gases and global warming.

tives, intent on validating the panel's original assumption that global warming is a man-made phenomenon. Indeed, two of its three working groups and an associated task force all focus on the impact and mitigation of global warming, based on the underlying presumption that it *is* a direct result of human activities. The other working group concentrates on the science, but shares the same biased assumption.

The IPCC's climate scientists consist of working scientists who act as either authors (contributors) or reviewers of the organization's reports. Writing and reviewing the 2007 Report involved more than 3,750 people,[18] of whom an estimated 2,000 are climate scientists according to press reports at the time. But of these, only a small percentage hold a PhD degree – the most generally accepted measure of scientific expertise.

In any case, the 2,000 is almost certainly an overestimate, and I couldn't come close to that number myself in a rough count of authors and reviewers listed on the IPCC website. So more than a half, perhaps as many as two thirds, of the 3,750 involved in producing the recent report are not climate scientists.

Many of those who are not climate scientists are not scientists at all: the academic participants listed by the IPCC include social scientists and geographers, and other contributors include civil engineers and even lawyers. This means that much of the scientific analysis in IPCC reports comes from people who simply do not have the background to assess the science. This is undoubtedly why there have been a number of reports of IPCC climate scientists whose dissenting opinions have been suppressed or ignored. It's not an approach that produces good science.

COMPUTER MODEL DELUSIONS

To skeptics, the IPCC's claim that we can be more than 90% certain that global warming is entirely man-made borders on the absurd. As we'll see, every single conclusion and prediction in all the IPCC reports is based on computer models of the Earth's climate, and these models are far from the tools for accurate climate analysis and prediction that the IPCC believes them to be.

I am not saying that computer models can't play a useful role in simulating complex physical phenomena such as climate. It's just that computer simulations are only as accurate as the underlying assumptions in the model.

If the assumptions in a particular computer model are based on established science, the predictions of the model are highly reliable – and a confidence level approaching 100% is warranted. A good example of this is the computer calculations used by the U.S. government to simulate explosions of nuclear weapons in its stockpile.

Back in the days when people worried about dangerous radioactive fallout from atmospheric testing of atomic bombs, worldwide political pressure led to the banning of atmospheric tests and the initiation of underground testing. But because some nuclear fallout occurs even when tests are carried out deep underground, computer simulations are now used as an alternative to conducting actual tests of the warheads. So accurate is the computer model in representing the science of nuclear explosions that the government believes it has an excellent handle on the capabilities of its nuclear stockpile, and consequently feels no need to resume its previous real-world testing to confirm the strength of its nuclear deterrent.

That's the good news, for both nuclear weapons stewardship and the environment. But not all computer models are as sophisticated as those that mimic nuclear explosions, nor are the underlying physical processes as well-understood or even known in many cases. Models that attempt to simulate the behavior of the universe, or of the human brain – both immensely complex systems, on different scales – are only in their infancy today and unable yet to make many useful predictions.

Climate models are in this category too. There's a lot we've learned about the intricacies of the climate on our planet, but there's also a great deal we don't know. Our understanding of clouds and water vapor, of the interaction between the oceans and the atmosphere, and even the details of the sun's effect on the Earth's climate are all still at a primitive stage. These factors can only be modeled crudely, and anything poorly understood is often left out of the models altogether.

It makes no sense at all, therefore, to attach a confidence level of 90% ("very likely" in IPCC terminology) to the statement, for example, that greenhouse gases cause global warming. That's still a possibility that we can't rule out for sure right now, but it's looking more and more *un*likely. A confidence level of a few percent might be closer to the mark.

THE SKEPTICS' EXPLANATION: NATURAL CAUSES

Skeptics about the alarmist belief in human-induced global warming have an alternative explanation that doesn't depend on bad science or demonization of critics to convey its message: the warming is predominantly natural, the result of one of nature's many cycles. We already know there are cycles that have caused the Earth's temperature to fluctuate numerous times in the past.

What are these cycles?

One type of cycle that is definitely *not* the cause of the current warming trend is regular but long-term changes in the Earth's orbit around the sun. We learn in high school that the Earth goes around the sun in an elliptical path, and that it spins on an axis that is tilted. Over time, the elliptical orbit stretches and contracts, the angle of tilt changes – on different time scales – and the Earth also wobbles on its axis, on yet another time scale.

The cumulative effect of all these slow dance moves by the Earth is that the amount of heat and light from the sun goes up and down over long periods of time, especially near the North Pole. This causes extended global warming and global cooling cycles, both of which can last for tens or even hundreds of thousands of years.

Prolonged cooling cycles are known as ice ages because of the massive ice sheets and glaciers that cover a lot of the planet. The next ice age is not expected for another 10,000 years or more, although the current global warming trend could alter that.

But there are other natural cycles, many of which are shorter than the Earth's orbital cycles, that could be influencing our present climate – notably those associated with our sun. And natural short-term climate cycles such as El Niño or the recently discovered Pacific Decadal Oscillation could have a greater impact on global temperatures than we think.

The sun, of course, is what makes life on Earth possible. Sunlight provides the energy that living organisms need to exist and grow.[19] The sun's heat energy, in combination with greenhouse gases (mostly water vapor) in the atmosphere, provides a sufficiently comfortable living environment for us to survive.

Like the Earth's orbit in the solar system, the sun's output is not constant but wiggles in time. However, variations in solar activity are very slight, and most solar cycles are much shorter in length than the 10,000-year-plus cycles that the

Earth goes through. For example, the number of sunspots[20] fluctuates over an interval of about 11 years. There are also longer cycles in solar output of approximately 85 years and 210 years.

The computer models that form the basis for the IPCC reports do incorporate solar effects, but only direct effects and only at a fairly rudimentary level. Missing from the models , as the IPCC readily admits,[21] are a number of indirect solar effects.

One indirect effect is the influence on the Earth's atmosphere of cosmic rays[22] that emanate from deep space and constantly bombard the atmosphere, sometimes creating low-level clouds that result in cooling of the Earth's surface. The sun can block these cosmic rays, changing the cooling effect.

Other indirect effects omitted from IPCC climate models include enhanced ultraviolet radiation from the sun at times of high solar output, which may heat the atmosphere by formation of ozone. The climate models are far from complete and have several other weaknesses in what they do include.

In fairness to the IPCC, it would be difficult at the present time to include indirect solar effects and other poorly understood physical mechanisms in the models, simply because our current understanding is so limited. But where I and other skeptics take issue with the IPCC is that, knowing that so much has been omitted from the climate models, the panel then goes on to draw conclusions on global warming that completely ignore the omissions and insists that most of the warming must come from man-made CO_2. And, as we'll see, the IPCC uses bad science to shore up its case.

Chapter 2: The Fuss About CO$_2$

Carbon dioxide gets a bad rap these days. Although crops and trees need it in order to grow, and it's the gas that gives soft drinks their fizz, we humans are spewing vast quantities of CO$_2$ into the air by burning coal and natural gas for energy and by driving cars. If you believe the global warming alarmists, all the CO$_2$ we produce is what's causing the mercury to rise around the world.

How could this be?

To answer that question, we need to examine the scientific observations on which the whole edifice of man-made global warming is founded. Observations, or data gathering, are at the root of the scientific method that I referred to earlier. But we'll see that these data don't necessarily mean global warming is a human phenomenon.

THE CO$_2$ GLOBAL WARMING HYPOTHESIS

There are two observations that have sparked the debate on global warming, one involving temperatures and the other, CO$_2$ levels:

1. Worldwide temperatures have been climbing since the middle of the 19th century, and

2. The amount of CO$_2$ in the atmosphere has increased over the same timespan. We're pretty sure that much of the extra CO$_2$ comes from human activities, mainly the burning of coal, oil and natural gas.

On top of this, the CO_2 level is increasing at an accelerating rate, as was the temperature, at least during the 1990s.

If temperature and the CO_2 concentration have both gone up (see Figure 2.1), then it's reasonable to say they must be connected – right? There is indeed a connection between rising temperatures and higher atmospheric CO_2, but this doesn't necessarily explain global warming. The crucial question is how much of a connection?

We know that rising CO_2 levels boost the Earth's temperature, but this is only a minuscule effect, not enough to account for the warming that we're seeing. It's entirely possible that higher temperatures are only weakly linked to the elevated CO_2 level and that most of the temperature increase doesn't come from CO_2, but from some other, natural source – many skeptics would say the sun.

The global warming debate hinges on this issue, on how the separate observations about the Earth's temperature and the CO_2 content of its atmosphere should be interpreted.

According to the IPCC and global warming alarmists, the only possible interpretation is that the warming we are now experiencing is caused by the raised level of CO_2. This conclusion is embodied in a scientific hypothesis (Table 2.1), which explains the CO_2–temperature connection in terms of what we call the greenhouse effect.

The greenhouse effect, named (though incorrectly) for the process that ripens tomatoes in a glass hothouse, is a well-understood scientific phenomenon. Because it's described at length in other global warming books and elsewhere, I won't go into detail here. A simplified explanation is that greenhouse gases in the atmosphere act as a radiative blanket around the Earth, trapping some of the sun's heat that would normally be radiated away.[27] This makes the planet warmer than it would be without greenhouse gases.

A possible greenhouse connection between increased CO_2 levels and higher temperatures was first proposed by the Swedish chemist Svante Arrhenius, at the end of the 19th century. He later hypothesized that human activity could result in global warming.[28] But it was not until the IPCC began publishing its climate reports in the 1990s that belief in the CO_2 hypothesis became widespread.

Contrary to popular opinion, however, the major greenhouse gas in the atmosphere is not CO_2, but water vapor (H_2O). Water vapor accounts for about 70%

of the Earth's natural greenhouse effect and water droplets in clouds for another 20%, while CO_2 contributes only a small percentage, between 4 and 8%, of the total. The other greenhouse gases are ozone, methane, nitrous oxide and chlorofluorocarbons (CFCs, the gases formerly used in aerosol cans and refrigerators), which all make even smaller contributions to greenhouse warming.

Figure 2.1: The Modern Temperature and CO₂ Record

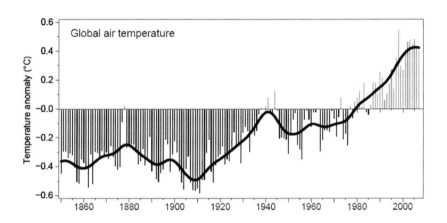

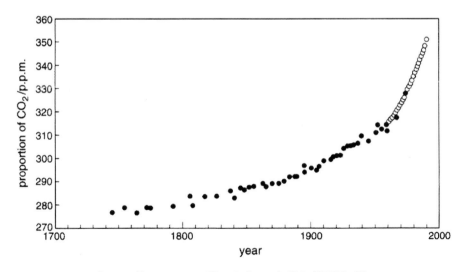

Sources: Temperature – Climatic Research Unit (CRU);[23] CO₂ (filled circles: Antarctic ice-core data, open circles: Mauna Loa Observatory measurements) – climateprediction.net.[24]

TABLE 2.1: THE CO_2 HYPOTHESIS

Observations

- Global surface temperatures have risen by approximately 0.8° Celsius (1.4° Fahrenheit) since 1850.[25]

- The CO_2 level in the lower atmosphere has gone up as much as 35% in the same period[26], largely due to man-made CO_2 emissions from factories and automobiles.

- Both temperatures[25] and CO_2 have been increasing more rapidly in recent years.

Hypothesis

Global warming is caused primarily by man-made CO_2 in the atmosphere, via the greenhouse effect, which says that greenhouse gases such as CO_2 heat up the Earth.

WHAT'S WRONG WITH THIS HYPOTHESIS?

1. *The CO_2 steady level problem:* The CO_2 level remained steady during previous global warming and cooling periods over the last 2,000 years – neither going up nor down as the average temperature rose and fell.

2. *The CO_2 amplification problem:* The climate change from extra CO_2 is very small. A 35% upswing in CO_2 causes only a tiny temperature increase, unless this increase is amplified by water vapor in the atmosphere and by clouds. But we don't know how big or small this amplification is, or even if it's an amplification and not a diminution.

3. *The CO_2 lag problem:* Historically, gains in atmospheric CO_2 levels occurred several hundred years *after* the temperature went up. This CO_2 lag, in the global warming period following an ice age, can't be reconciled with today's global warming, in which CO_2 and temperature have risen together.

Without the natural greenhouse effect – in the absence of any greenhouse gases at all – life on Earth as we know it would not exist. The globe would be cooler than it is now by about 33° Celsius (60° Fahrenheit), too chilly for most living organisms to survive.

Thus, it's definitely possible in principle that adding to the store of existing greenhouse gases by putting more CO_2 into the atmosphere could increase temperatures. One place in the solar system where there is an abundance of CO_2 and a pronounced greenhouse effect is the planet Venus.[29] But the total amount of CO_2 in the Earth's atmosphere is still only very small: about 390 parts per million, or less than one twentieth of a percent. It takes an awful lot of CO_2 to make an appreciable difference.

The critical issue for us is whether the extra CO_2 that we've injected into the Earth's atmosphere since the early 1800s is enough to bump up the temperature by the observed 0.8° Celsius (1.4° Fahrenheit).

It is on this question that global warming skeptics and alarmists differ most sharply. Skeptics say there's no good evidence that man-made CO_2 is playing any significant role in global warming. The alarmists, on the other hand, say everything points to the fact that it is.

RED FLAGS FOR THE CO$_2$ THEORY

Despite what alarmists believe, the CO_2 hypothesis is hardly a satisfactory explanation for global warming. There are several red flags – listed in the "What's Wrong with this Hypothesis?" section of Table 2.1 – which present big problems for the hypothesis. Here we briefly review these three scientific roadblocks for the man-made CO_2 theory, though we won't delve into them more fully until later.

The *CO$_2$ steady level problem* has to do with attempts to match CO_2 and temperature records over the last 2,000 years. If CO_2 is responsible for current global warming, as the hypothesis states, then past warm periods should have been accompanied by higher CO_2 levels also, and cool periods by lower CO_2 levels. That's certainly what happened hundreds of thousands of years ago, during the ice ages, when ice-core data reveal that temperature and CO_2 marched up and down in lockstep.

But if we look at recent CO_2 and temperature data, that's not what we see. For the last 2,000 years the CO_2 level has been almost flat as a board (see Figure 3.2), but this was during an era when the temperature bounced around and there were several documented warming and cooling periods. The *only* recent time when the CO_2 concentration has changed along with temperature is the modern warming period since 1850. The CO_2 hypothesis can't explain this difference between the modern and pre-modern behavior of the CO_2 level.

The second red flag for the CO_2 theory is what I've called *the CO_2 amplification problem*. This refers to the sensitivity of the climate system to CO_2 – how much does the temperature go up when we put a particular amount of CO_2 into the atmosphere?

If we consider just the natural greenhouse effect, the present increase in atmospheric CO_2 from human activity, which may be as high as 35-40%, barely affects global temperatures. That is, the climate sensitivity for CO_2 on its own is rather small. Even a doubling (a 100% increase) of the CO_2 level would only raise the temperature by about 1° Celsius (1.8° Fahrenheit)[30] – or even less than that[31] – and we're nowhere near doubled CO_2 yet.

Carbon dioxide can only be the global warming culprit if the Earth's natural greenhouse effect for CO_2 is amplified somehow. In the IPCC's climate models, the tiny effect from CO_2 acting alone is magnified by water vapor in the atmosphere and by clouds, through what are known as feedback mechanisms. In other words, the presence of water vapor and clouds makes the atmosphere much more sensitive to small additions of CO_2 than it would be otherwise.

The problem with this picture is that we don't know if it's correct. We're not sure whether water vapor and cloud amplification make a big contribution to the temperature or a little one, or even if they boost the temperature at all. Is the effect of CO_2 more like a puff of wind or a 50 mph gale? We'll return to this question in Chapters 4 and 5.

The third red flag for the CO_2 global warming hypothesis is *the CO_2 lag problem*. This refers to the time lag during past ice ages between the CO_2 level in the atmosphere and the temperature, both deduced from measurements of air trapped in ice cores. The data shows that rises and falls in CO_2 occurred 600 to 800 years after the corresponding temperature changes.

That doesn't jibe with the nearly simultaneous increases in temperature and CO_2 level, as proposed in the CO_2 hypothesis to explain current global warming – more on this later also.

Chapter 3: Science Gone Wrong

If you believe what it says, the IPCC started out with honorable intentions, despite the bias toward man-made global warming in its original charter. In discussing the nature of science in the historical overview of its 2007 report, the IPCC states:

> Science may be stimulated by argument and debate, but it generally advances through formulating hypotheses clearly and testing them objectively. This testing is the key to science.
> ... Thus science is inherently self-correcting; incorrect or incomplete scientific concepts ultimately do not survive repeated testing against observations of nature. [32]

This is an accurate and succinct summary of what good science *should* be all about – the essence of the scientific method[33] that I referred to earlier.

But the IPCC then goes on to say:

> Using traditional approaches, unequivocal attribution of causes of climate change would require controlled experimentation with our climate system. However, with no spare Earth with which to experiment, attribution of anthropogenic climate change must be pursued by ... demonstrating that the detected change is consistent with computer model simulations ...[34]

"Must" be pursued by computer simulations? Although the IPCC correctly acknowledges that it's impossible to test the CO_2 hypothesis by conducting controlled experiments on our Earth, its insistence that the theory of human-induced global warming can be validated by computer modeling is where the bad science begins.

Unfortunately for science, the IPCC missteps go on. It's not just blind faith in computer models that derails the IPCC's conclusions, but a host of other departures from sound scientific practice as well – including outright fabrication.

DATA MANIPULATION

Central to any scientific investigation, such as checking out the validity of the CO_2 global warming hypothesis, are the raw data gathered by observation. Without data there can be no hypotheses, no science.

And the data must be handled according to certain unwritten rules, if inferences drawn from the data are to be regarded as good, solid science. These rules include examining all the evidence and not discarding anything that doesn't seem to fit, eliminating bias[35] in the measurements, and using multiple sources of data to minimize the influence of any personal quirks of the investigators .

The rules are really just common sense, but important nonetheless because science strives to understand the physical world through honest investigation and discovery. We can't hope to gain any insight if we play fast and loose with the data.

Regrettably, the IPCC does just that with the two principal pillars of its data edifice – the temperature data and the CO_2 record. And not just once, but several times over.

Exaggerated Temperatures

A big part of the IPCC story is the steep increase in global temperatures since 1980, which is clearly visible if you look back at Figure 2.1. That data, measuring the temperature anomaly – or change from the average temperature – for the period from 1850 to 2005, is based on both land and sea measurements.

But how accurate are these measurements?

I'm not talking about the thermometers used, since thermometers have been around for a long time and we can depend on them to accurately record the temperature. But that doesn't necessarily mean that temperature readings are reliable. The reading will depend on where the thermometer is placed – a thermometer out in the hot sun will show a different temperature from one nearby in the shade, for example.

So land-based surface temperatures are always taken by thermometers in special white louvered boxes, about three feet off the ground. This standardizes the measurement method, but there's still a problem because of what climatologists call the urban heat island effect.

The term urban heat island refers to the warming generated by people living in urban communities, which are always significantly warmer than surrounding rural areas because of the tendency of concrete, asphalt and buildings to absorb heat. Heat islands introduce bias into temperatures that are averaged over both city and rural land areas, causing the temperatures to be overstated.

Temperature bias can also arise because the white louvered box is in the wrong place. If it's next to an asphalt parking lot, for instance, heat reflected by the asphalt and heat generated by the running engines of vehicles will skew the measured temperature to indicate a warming that is not real. Regulations in the U.S. require temperature sensors to be at least 100 meters from artificial heating or reflecting surfaces, but many temperature stations don't meet this requirement.

The urban heat island effect is most pronounced in the largest cities, but has been observed in smaller centers as well. In the U.S. state of Illinois, a study revealed that average soil temperatures at a rural site rose by 0.4° Celsius (0.7° Fahrenheit) between 1882 and 1952. Over the same period, the surface air temperature from three nearby small towns with less than 2,000 inhabitants showed an increase of 0.57° Celsius (1.0° Fahrenheit).[36] The warming rate was significantly higher, by 43%, due to urbanization.

Climate scientists are well aware of the heat island effect, and over the years have developed an "urbanization adjustment" to correct recorded temperatures for the data bias that urbanization produces, a bias sometimes referred to as data contamination by human activity.

However, the IPCC, in its ongoing quest to make all its data conform to the CO_2 global warming hypothesis (Table 2.1), essentially ignores this data contamination. There is of course irony here: on the one hand, the IPCC invokes human industrial activity to explain global warming but, on the other hand, rejects evidence for the influence of man-made cities on the temperature!

Economist Ross McKitrick and climatologist Patrick Michaels have studied the pattern of warming over the Earth's land surface compared to local economic conditions, which are a signature of the human presence. In a comprehensive statistical and economic analysis of global temperature data, they conclude that the probability that human activities such as industrialization and urbanization do *not* influence local temperature trends is less than 1 in 14 trillion.[37] That's a staggering number, and means that urbanization unquestionably creates a net warming bias.

McKitrick and Michaels find that recently measured global warming rates need to be lowered appreciably to cancel out this bias, even though the warming rates have supposedly been adjusted for urbanization already. When properly corrected for the urban heat island effect, the warming rate recorded on land since 1980 falls by about half.

A similar conclusion was reached earlier by two Dutch meteorologists, Jos de Laat and Ahilleas Maurellis, using different data and a different testing methodology (part of which involved using an IPCC climate model).[38] They showed that a statistically significant correlation exists between the geographic patterns of warming and industrial development, a correlation that is not simulated by IPCC climate models. Like McKitrick and Michaels, they found that the correlation biases the global warming trend upward, so that measured temperatures must be corrected downward.

Needless to say, the IPCC dismisses both studies in its 2007 Fourth Assessment Report. The IPCC claims that the heat island urbanization adjustment is negligible and that there is hardly any bias in uncorrected temperature trends, quoting studies by other climatologists.[39]

This has led to McKitrick – who was an external reviewer for the report, and submitted extensive comments critical of the IPCC position – leveling charges of fabrication against the IPCC, which I'll discuss later in the chapter.

Using McKitrick and Michaels' result that the land surface warming rate since 1980 should be cut in half because of urbanization, the corresponding drop in the post-1980 *global* warming rate is about one fourth (25%).[40] The correction to global warming is smaller than for land regions alone, since the oceans are warming at a slower rate and show no heat island effect.

Because urbanization goes all the way back to the 19th century – although it has accelerated in recent years – it is highly probable that the global warming rate for the whole period since 1850 has been overestimated by the same amount, and not just the warming rate from 1980. What this means is that the IPCC's estimated temperature increase of 0.8° Celsius (1.4° Fahrenheit) for the modern period is too high and should be reduced by one fourth, to 0.6° Celsius (1.1° Fahrenheit).

An exaggeration of one third[41] in the global temperature increase may not seem like a big deal, but it unquestionably is. The IPCC argument that natural variability alone cannot explain the observed rise in worldwide temperatures becomes a lot weaker if that rise has been overestimated by a third, or 0.2° Celsius (0.4° Fahrenheit), not to mention that the IPCC's climate models then have much less validity.

Some skeptics maintain that the temperature overestimate is more than one third. But even at a third, the exaggeration is, at the very least, poor science – poor science that has led the IPCC to make numerous unjustifiable predictions of disastrous consequences of global warming that await the Earth. Good science demands intellectual honesty, including correction of data for bias.

More evidence for bias in land temperatures comes from satellite data. Satellites in orbit around the Earth can measure temperature accurately over both land and sea, with the exception of small regions near the North and South Poles, by means of microwaves. These measurements are also subject to bias, caused by the satellite drifting in its orbit and other factors, but all these factors are well understood and can easily be corrected for.

Satellites sample global temperatures uniformly, unlike Earth-based thermometers that are weighted more toward developed, urban areas. Taking the heat island effect into account, the satellite data therefore shows less warming than the land surface records relied on by the IPCC. This is no doubt why the IPCC has chosen not to include satellite temperature measurements in its estimate of the recent global warming rate, even though data is available from 1979 (Figure 3.1).

In fact, the land warming rate from the satellite measurements is almost as low as the heat-island corrected warming rate, based on thermometers, of McKitrick and Michaels.[37] So either the satellite data are wrong, which no one – alarmists or skeptics alike – believes, or urban contamination causes bias in the surface data, bias that the IPCC ignores.

A further piece of evidence that the IPCC has exaggerated the rise in global temperatures, especially over the period since 1980, is the sudden drop in the number of temperature measuring stations that occurred in the 1990s – from around 5,000 to 2,000 or so.[44] Many of these stations were in cold parts of the globe such as Arctic regions of the former Soviet Union, which had collapsed in 1989, prompting the shutting down of remote weather stations that were too expensive to maintain.

Figure 3.1: The Satellite Temperature Record

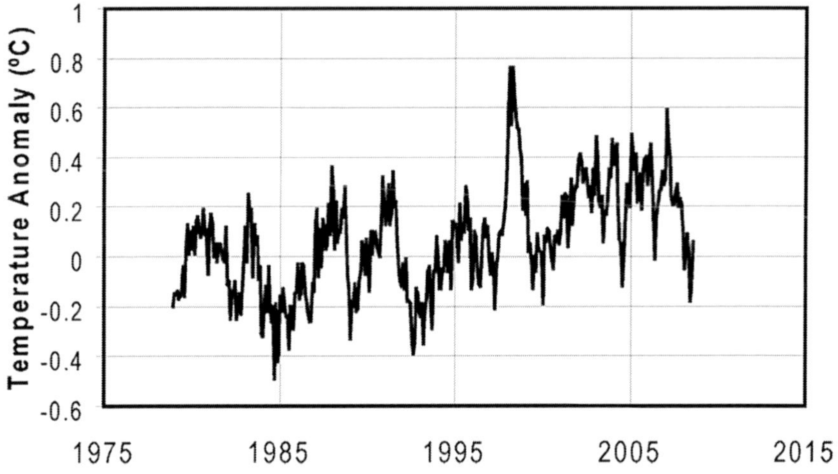

Source: Adapted from Hayden;[42] temperature data from the University of Alabama in Huntsville.[43]

The abrupt loss of stations in cold locations would almost certainly have raised the average temperature measured by the remaining stations. Yet this issue has never been addressed by the IPCC, whose intention is to prove by any means possible that global warming comes from human activity. Since the atmospheric CO_2 level has been going up at an escalating rate in recent years, the temperature must have been doing the same thing if the CO_2 hypothesis is to explain the warming. So for the IPCC, any inflation of recent temperatures is warranted, whatever the evidence to the contrary.

Disappearing Temperatures: The Hockey Stick

Much worse than a one third exaggeration in the modern temperature increase was the "hockey stick" scandal – an outrageous attempt by the IPCC to skew historical temperature data to suit its political agenda. The episode is well-documented elsewhere but bears repeating here.

The scandal arose because of the IPCC's need to validate its hypothesis about the connection between global warming and man-made CO_2. We've seen how this hypothesis is based on similar upward trends in the modern temperature record and the CO_2 level (Figure 2.1). If the hypothesis, and computer climate models that depend on the hypothesis, are to hold up, then temperature and CO_2 should track one another historically and not just for the last 160 years.

The difficulty with this need is that, over the last 2,000 years, the temperature and CO_2 level *don't* track (Figure 3.2). The temperature has fluctuated, both up and down, but there has been almost no change in the CO_2 concentration until modern times – the CO_2 steady level problem that I referred to in the previous chapter.

How was this historical data obtained?

Measurement of temperature using modern scientific thermometers goes back only to the early 18th century, and accurate determination of the CO_2 level has been possible only for the last 50 years or so. Temperature and CO_2 data for earlier periods come from so-called proxy methods, or indirect measurements using sources such as tree rings, ice cores, leaf fossils, or boreholes.

Each of these proxy methods has its limitations. Although the most commonly used proxy for temperature is tree-ring data, some paleoclimatologists (climatologists who study the past) believe that tree rings are unreliable indicators. This is because the widths of tree rings respond not only to temperature, but also to other factors such as moisture and CO_2. However, the data in Figure 3.2 were not based on tree rings.

The distinctly noticeable warm spell seen around the year 1000 is known to historians as the Medieval Warm Period, a time when warmer than normal conditions were reported in many parts of the world. The cool period centered around the year 1650 has been labeled the Little Ice Age and is also reported in various historical records. But there is no sign at all of these warming and cooling periods in the CO_2 data for the same timespan, which is based on ice-core proxies.

As I said, this mismatch is a problem for the IPCC's view of global warming. For the CO_2 hypothesis to be correct, the temperature and CO_2 level must go hand in hand, for *all* periods of time including the last 2,000 years.

Oddly enough, the IPCC seemed unaware of this problem in its Second Assessment Report in 1995, which showed a temperature graph for the last 1,000 years with both the Medieval Warm Period and the Little Ice Age not only included, but clearly labeled (Figure 3.3).

Figure 3.2: The Last 2,000 Years of Temperature and CO_2

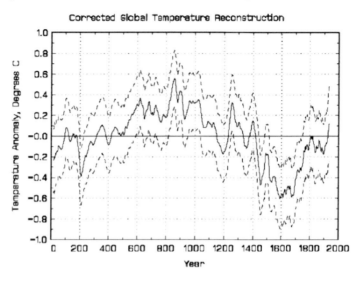

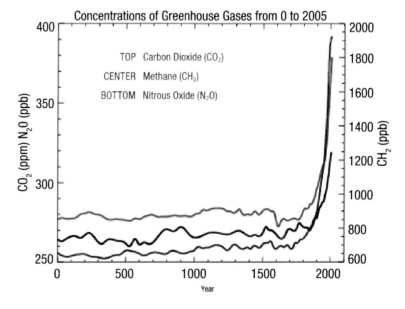

Sources: Temperature (showing 95% confidence intervals) – Loehle and McCulloch;[45] Greenhouse gases – Intergovernmental Panel on Climate Change.[46]

Figure 3.3: The IPCC's View of History, 1995

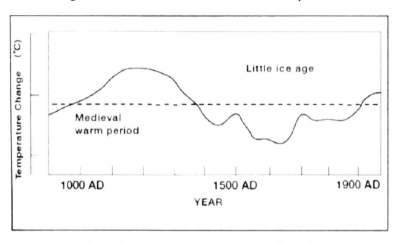

Source: Intergovernmental Panel on Climate Change.[49]

Figure 3.4: The IPCC's View of History, 2001 – The "Hockey Stick"

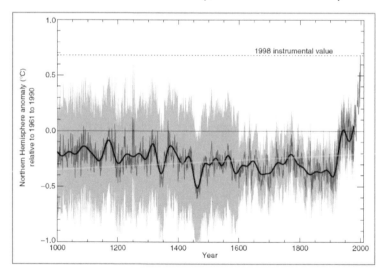

Source: Intergovernmental Panel on Climate Change.[50] Note that the horizontal time scale lines up approximately with Figure 3.3 above.

Yet the Third Assessment Report in 2001 told a radically different story. All of a sudden, the Medieval Warm Period and the Little Ice Age had disappeared! In their place was a fairly flat-looking graph (Figure 3.4) with few temperature ups and downs until the beginning of the present climb around 1900 – a chart that now bore a remarkable resemblance to the modern CO_2 record, having the shape of a hockey stick on its side.

Hey presto! At a stroke, the IPCC solved its problem. The temperature record for the past 2,000 years indeed showed the same behavior as the CO_2 level (and that of other greenhouse gases), and the IPCC could now proclaim that it was right about global warming being human-induced. If the panel was rewriting history at the same time, so be it.

The hockey stick graph was the work of a group led by Michael Mann, an IPCC author then at the University of Virginia, who in 1998 and 1999 reconstructed historical temperatures for the period from 1000 to 1900, using tree-ring data almost exclusively. Mann and his group then grafted the 20th century thermometer record onto this pre-1900 proxy record.[47]

Never mind that tree rings are considered an inaccurate proxy for past temperatures, nor that the 20th century record is exaggerated by the urban heat island effect, as we've just seen. The Mann graph had an immediate visual and political impact. By doing away with the Medieval Warm Period and the Little Ice Age, the hockey stick not only vindicated what global warming alarmists had been saying, it also gave a boost to governments wavering on adoption of the UN's 1997 Kyoto Protocol, which limits emissions of CO_2.

Yet this apparent triumph for the IPCC's global warming model was about to come crashing down around its ears, with the subsequent revelation that – yes, you guessed it – the IPCC was guilty of egregious data manipulation, of bending scientific data for an ulterior motive.

The Mann studies and the hockey stick were initially debunked in 2003 by Canadian statistician Stephen McIntyre and economist Ross McKitrick (co-author of the urban heat island studies discussed earlier), who found that Mann's conclusions were based on faulty statistical analysis as well as other errors.[48] In fact, McIntyre and McKitrick showed that they could almost always produce a hockey stick, even from completely meaningless random data. In their words,

The particular "hockey stick" shape derived in the Mann proxy construction ... is primarily an artefact of poor data handling, obsolete data and incorrect calculation of principal components.[51]

The authors added that Mann's studies were heavily dependent on a small subset of the tree-ring data – ring widths from North American bristlecone pines, which are widely doubted to be reliable temperature proxies because of an unexplained 20th century growth spurt. Omission of this bristlecone pine subset, representing just one of over 100 data sets included in the original analysis, reinstates medieval warming and gives the lie to the IPCC's assertion that our present warm trend is exceptional compared to preceding centuries.

In 2006, some five years after the publication of the IPCC's report featuring the hockey stick, a team of statisticians appointed by the U.S. House Committee on Energy and Commerce found Mann's statistical analysis to be "somewhat obscure and incomplete", and the criticisms by McIntyre and McKitrick to be "valid and compelling".[52] The team also accused the IPCC of politicizing Mann's work.

At almost the same time, the U.S. House Committee on Science, which had been charged by the National Research Council (NRC) of the National Academy of Science to report on temperature data for the last 2,000 years, came to similar conclusions. The NRC report states:

Large-scale surface temperature reconstructions yield a generally consistent picture of temperature trends during the preceding millennium, including relatively warm conditions centered around A.D. 1000 (identified by some as the "Medieval Warm Period") and a relatively cold period (or "Little Ice Age") centered around 1700. The existence and extent of a Little Ice Age from roughly 1500 to 1850 is supported by a wide variety of evidence including ice cores, tree rings, borehole temperatures, glacier length records, and historical documents. Evidence for regional warmth during medieval times can be found in a diverse but more limited set of records including ice cores, tree rings, marine sediments, and historical sources ...[53]

Despite this widespread denouncement of his work, Mann – who is a paleoclimatologist but not a statistician – has continued to argue for the legitimacy of the hockey stick graph. At one stage, he defended the absence of the Medieval Warm Period and the Little Ice Age from his temperature reconstruction by saying that these were local rather than global phenomena, and restricted to small regions of the Northern Hemisphere. The difficulty with this explanation is that there is ample historical evidence from around the world, including the Southern Hemisphere, of the existence of both climate periods.[54]

In 2008, Mann's group published a new study reconstructing temperatures back to the year 700,[55] but without the use of any of the tree-ring data that formed the basis of their previous studies for the period since 1000. Although their latest paper concedes the occurrence of at least a slight medieval warming, Mann claimed in an interview that, far from being broken, "the hockey stick is alive and well".[56]

However, McIntyre and McKitrick are already on the case and early indications are that the new Mann study may be just as flawed in its statistical analysis as the 1998 and 1999 papers.

What the IPCC will make of Mann's recent work remains to be seen. Its 2007 Fourth Assessment Report grudgingly conceded that the hockey stick graph in the 2001 report was controversial, and that a more careful reconstruction of the temperature record does indeed show medieval warmth and cooler conditions during the Little Ice Age.[57] But an article published in 2005 by a University of Oklahoma geoscientist, David Deming,[58] leaves no doubt that in 2001, the IPCC was exploiting the hockey stick for its own ends.

Deming had established credibility with alarmists in the climate science community with an earlier paper, in which his analysis of borehole temperature data appeared to support the IPCC's CO_2 theory of global warming, although he concluded that natural variability could not be ruled out as a cause of warming either. The research was enough, nevertheless, to gain Deming admission to the alarmist club:

> With the publication of the article in Science ... They thought
> I was one of them, someone who would pervert science in
> the service of social and political causes. So one of them
> let his guard down. A major person working in the area of

climate change and global warming sent me an astonishing email that said "We have to get rid of the Medieval Warm Period."[59]

As we've seen, elimination of the Medieval Warm Period and the Little Ice Age was essential if the IPCC was to match up the temperature record and the CO_2 level over the last 2,000 years (Figure 3.2), and thus confirm its hypothesis that global warming comes from human activity. So Mann's hockey stick curve, erroneous and deceptive as it is, must have seemed like a gift from God.

Luckily for science, the world woke up to this particular case of IPCC corruption and the hockey stick is now largely discredited.

Cherry-picking the Data: Soaring CO_2

One of the rules of the scientific method, and one of the most frequently overlooked, is that you have to consider *all* the data. What it's saying is that you can't ignore any piece of evidence that doesn't fit your theory or verify your hypothesis, simply because it's inconvenient (Al Gore, please take note!).

I've done scientific research, and I know how tempting it can be to reject data that you don't like for some reason. Maybe you made an observation that conflicts with what everyone else has seen, or maybe you can't draw the trend line that you want through your graphed data points without throwing out some data.

But it's a big no-no in scientific circles to ignore or discard *any* experimental observation, unless there was an obvious mistake that calls for repeating the measurement, or there is bias in the data that can't be corrected for – which is sometimes the case with historical data. All other data is kept, even if it can't be fully explained.

Unfortunately, the IPCC has been very selective about the data it uses to reinforce its contention that global warming is a man-made phenomenon. This cherry-picking can be found in many areas, of which the available CO_2 data is a prime example.

Recall that the CO_2 hypothesis, stating that global warming is a direct consequence of human activity, links rising trends in worldwide temperatures and the level of CO_2 in the atmosphere. We've just seen how the IPCC tried to flatten out the bumps in the historical temperature record to make it conform to the shape of the CO_2 record, leading to the infamous hockey stick curve. The IPCC has also been less than honest in its use of historical CO_2 data.

The problem this time for the IPCC was not the shape of the CO_2 graph for the last 2,000 years (Figure 3.2), but rather the CO_2 level itself over the flat or steady level portion of the chart – for the period up to approximately 1850, when the CO_2 concentration began its present ascent due to industrialization. If the steady, preindustrial CO_2 level was higher than the IPCC says it is, which is about 280 parts per million,[60] the increase today isn't as large as we think and the IPCC is exaggerating again.

Most of the CO_2 data shown in Figure 3.2 and Figure 2.1 (illustrating just the modern period) is derived from ice-core proxies, but this is not the only data available. Among other observations, there is a large set of measurements of CO_2 levels in the Northern Hemisphere between 1812 and 1961, based on chemical analysis (Figure 3.5).

Figure 3.5: CO_2 Data Ignored by the IPCC

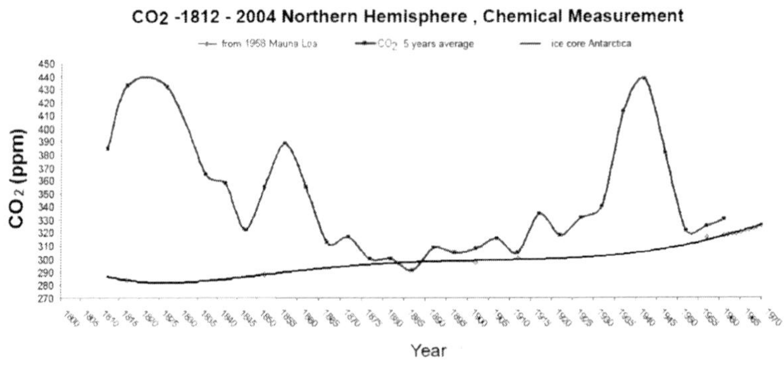

CO2 -1812 - 2004 Northern Hemisphere , Chemical Measurement

Source: Adapted from Beck.[61] The Antarctic ice-core data at the bottom of the graph is similar to that shown in Figure 2.1.

It's immediately obvious that the chemical measurements show a totally different trend than the ice-core proxy data, which are derived from modern optical measurements and are included in the same chart for comparison. Where the ice-core numbers depict a smooth, slowly rising CO_2 level, the chemical data show enormous swings over the same period.[62] And the average CO_2 concentration based on the chemical data, which is 321 parts per million for the period up to

1900, is much higher than the IPCC's preindustrial level of 280 parts per million deduced from ice cores.

Which set of data is right?

It appears at first glance that the smooth ice-core curve for CO_2 in Figure 3.5 is a better match to the underlying trend in the temperature record (Figure 2.1) than are the CO_2 data derived from chemical methods.[63] This is why the IPCC, in order to bolster its hypothesis about CO_2 and global warming, saw fit to completely ignore the chemical measurements – at least 70,000 of them over 150 years – and cite only data from ice cores in its reports, barely mentioning that any other CO_2 data even existed.

However, it's not just the IPCC that has ignored the chemical data in Figure 3.5. Back in 1939, British engineer Guy Callendar had arbitrarily selected a very small subset of these chemical CO_2 measurements, in order to support his claim that the CO_2 level had increased since the 19th century, causing temperatures to rise[64] – Callendar being an early believer in the CO_2 hypothesis.

Essentially, Callendar rejected all the chemical observations except the low-lying numbers that, like the later ice-core data, show a smoothly rising CO_2 level. But the statistical validity of this data selection was subsequently questioned,[65] in an eerie precursor of the recent debunking of the hockey stick. While the critic conceded that the CO_2 cherry-picking didn't refute Callendar's claim about CO_2 and temperature, he pointed out that the available data did not confirm it either.[66]

About the only reason that Callendar's highly selective choice of CO_2 data, endorsed years afterwards by the IPCC, has survived the test of time is that it appears to match the ice-core data so well. This may be no more than pure luck, since ice-core measurements of the CO_2 level are not as accurate as we think.[67]

In arriving at its preindustrial CO_2 level of 280 parts per million, the IPCC has chosen to disregard not only the chemical measurements from the 19th and 20th centuries, but also earlier data on the size of stomata – the small openings on tree leaves, which become bigger and smaller as the amount of CO_2 in the atmosphere varies.

Although some stomatal measurements are consistent with ice-core records,[68] other data is not and indicates a preindustrial CO_2 concentration as high as 340 parts per million,[69] comparable to the level derived from the chemical data in Figure 3.5.

Some climatologists dismiss all historical proxy data, including ice cores, for both CO_2 and temperature, saying that the methods used to obtain the data are indirect and unreliable. This is a valid scientific position to take, but the problem it creates for the debate on global warming is that lack of proxies leaves us with insufficient information to draw any conclusions at present. My view is that proxies can be useful, but they shouldn't be elevated to a higher level of importance than direct measurements such as the chemical data I've just discussed.

And just as temperature measurements are subject to bias from the urban heat island effect, background CO_2 levels can be contaminated by the presence of nearby sources of CO_2, both natural and man-made. For this reason, CO_2 data is always taken from locations well away from forests and other vegetation, which temporarily generate CO_2 from overnight respiration, and far from any industrial activity.

At least the IPCC takes care to eliminate bias from its CO_2 data, if not its temperature measurements.

One observation accepted by both alarmists and skeptics is that the current CO_2 background level is around 390 parts per million, a number based on highly accurate modern technology, and is continuing to go up.[70] But the important scientific question is whether this level has soared almost 40% from a preindustrial baseline of 280 parts per million, as the IPCC insists, or has risen by a lesser amount from a higher baseline of 320 parts per million or more.

Good science looks at all the data. If we do that, the increase in atmospheric CO_2 from 320 to 390 parts per million since 1850 is only about half as much (22%) as the cherry-picking IPCC has told us, based solely on ice-core data. This in turn means that the IPCC's estimate of the temperature boost from man-made CO_2 needs to be lowered – and that natural variability can explain most, if not all, of the global warming that has occurred.

We've now looked at three separate examples of how the IPCC has abandoned any pretense of playing by the scientific rules in arriving at its position that humans have caused global warming – by ignoring bias in the modern temperature record, trying to distort historical temperatures, and being highly selective

TABLE 3.1: BAD IPCC SCIENCE – DATA MANIPULATION

The IPCC take on climate data

- Global warming is about 0.8° Celsius (1.4° Fahrenheit) since 1850.

- The warming rate since 1980 is higher than ever before.

- Temperature and the CO_2 level in the atmosphere have always gone hand in hand, for all periods of time – past as well as present. This is required by the CO_2 hypothesis.

- The amount of CO_2 in the atmosphere has risen almost 40% from its preindustrial baseline level.

WHAT'S WRONG WITH THIS DATA?

1. Global warming is *exaggerated* by around one third, or 0.2° Celsius (0.4° Fahrenheit), because the IPCC has ignored a warming bias caused by artificially high temperatures measured in urban areas. The bias is consistent with satellite data showing lower warming.

2. The IPCC has *inflated* the recent warming rate, both by ignoring bias and by not accounting for the 1990s closing of weather stations in cold parts of the world.

3. To match temperature to the CO_2 level over the last 2,000 years, the IPCC *rewrote history* by eliminating the well-established Medieval Warm Period and the Little Ice Age, creating the erroneous hockey stick graph.

4. To match up the temperature and CO_2 records since 1850, the IPCC has *cherry-picked* the available CO_2 data so as to exaggerate the rise in CO_2. The actual gain in CO_2 from preindustrial times is only about half as much as the IPCC says.

about its use of CO_2 data. If submitted as part of a science thesis by a PhD student in a reputable institution, any one of these efforts alone would be enough to fail the student.

But there's more. We'll see in the next section that the IPCC not only thumbs its nose at accepted procedures for gathering scientific data, but it also stoops to shady and corrupt practices in presenting and publishing that data.

CORRUPTION

Biased from the beginning toward its belief in man-made global warming, the IPCC – aided by its alarmist allies in the public – has spared no effort in attempting to suppress contrary scientific evidence and to stifle the views of critics. This line of attack has extended to blatant dishonesty as we've just seen, and even to making fraudulent claims.

Cooking up Evidence

As discussed earlier in this chapter, the IPCC's 2007 report minimizes the influence of urbanization on recorded temperature data. Imagine its surprise, then, when upstarts McKitrick and Michaels published their detailed statistical study showing that the urban heat island effect accounts for just over a half of the measured temperature increase on land due to global warming![71]

Perhaps in the hope of silencing any public comment by McKitrick, the IPCC had made him an external reviewer for that part of the report. Their efforts backfired, however, as McKitrick not only submitted extensive comments in support of the conclusions he and Michaels had reached, but also struck back openly at the IPCC when those comments were passed over.[72]

In explaining why he accepted the IPCC's invitation to serve as a reviewer, McKitrick says:

> Contamination of surface climate data is a potentially serious problem for the IPCC. Conclusions about the amount of global warming, and the role of greenhouse gases, are based on the assumption that the adjustment models work perfectly ... the core message of the IPCC hinges on the assumption that their main surface climate data set is uncontaminated. And by the time they began writing the

recent Fourth Assessment Report, they had before them a set of papers proving the data are contaminated.[73]

The IPCC report dismisses the McKitrick and Michaels study with the false statement that most global temperature measurements have already been adequately adjusted for urbanization, and that only a very small additional correction is necessary. McKitrick had written a lengthy criticism of this opinion, backed up by strong evidence to the contrary from the study, during the writing of the report.

What happened next is astounding, but typical of how the IPCC operates. Confronted with published evidence from one of their own reviewers that the heat island effect contaminates global temperatures, they trivialized the evidence with the conjecture that urban warming can be chalked up to entirely natural causes. In publicly available comments on the report's second draft, the IPCC authors state that the urban heat island effect is insignificant and that any urban warming comes from "...strengthening of the Arctic Oscillation and the greater sensitivity of land than ocean to greenhouse forcing owing to the smaller thermal capacity of land."[74]

This preposterous claim is complete nonsense. The Arctic Oscillation is a wind circulation pattern that affects long-term weather trends in the Arctic, but has absolutely nothing to do with the heat island effect from urban areas. And the comparison of land to ocean is irrelevant since McKitrick and Michaels only studied temperatures over land.

But the IPCC didn't stop there. Referring to the strong connection that McKitrick and Michaels found between temperatures and local economic activity, the final version of the report declares:

> However, the locations of greatest socioeconomic development are also those that have been most warmed by atmospheric circulation changes (Sections 3.2.2.7 and 3.6.4), which exhibit large-scale coherence. Hence, the correlation of warming with industrial and socioeconomic development ceases to be statistically significant.[39]

Here the IPCC acknowledges the correlation between warming trends and socioeconomic development, meaning industrialization and urbanization, but dismisses the correlation as a mere coincidence due to unspecified "atmospheric circulation changes". This is more of the Arctic Oscillation gobbledygook that the IPCC authors invoked earlier in the report writing process, since the two cited sections of the report say nothing at all about industrial or urban development.

However, the most outrageous comment is the second sentence in the excerpt above, maintaining that McKitrick and Michaels' evidence for a definite correlation between elevated temperatures and urbanization is statistically insignificant. McKitrick has since demonstrated in a further study, which specifically includes four major atmospheric circulatory (wind) patterns, that exactly the opposite is true – that is, the correlation is indeed statistically significant, and measured temperatures are strongly tied to urbanization.[75]

The IPCC's claim to the contrary is a flagrant abuse of science and scientific methodology. Not only did the UN panel reject perfectly sound scientific evidence in McKitrick and Michaels' previous studies, but they did so on the basis of nonexistent counterevidence from wind patterns. In McKitrick's words, this amounts to "making stuff up" and constitutes a "plain fabrication" by the IPCC.[72]

Data Fraud

Creating fictional evidence to prop up a fallacious argument about urban heat islands is not an isolated example of fraud by the IPCC. Another instance that has come to light recently, also involving the effect of urbanization on global temperatures, concerns fabrication of the actual raw data – although this time the alleged perpetrator is an IPCC climate scientist, rather than the IPCC itself.

The transgression is sufficiently serious that it has led to a formal allegation of fraud against the scientist, Wei-Chyung Wang, who is a professor at the State University of New York at Albany. The allegation was originally made to the university in August 2007 by UK mathematics researcher Douglas Keenan. Following an internal university investigation that claimed to find no evidence of fraud, Keenan accused the university of "the appearance of a cover-up" and took the case to the State Attorney General's Office, renewing the original allegation in July 2008.[76] The complaint is currently under review.

The alleged fraud concerns two 1990 research papers co-authored by Wang, in which a comparison was made between temperatures gathered at both urban and rural weather stations in China over the years 1954–1983. Wang was the lead author of one paper;[77] he contributed the same Chinese data to the other paper, which also included temperature data from rural areas in the former Soviet Union and Australia.[78]

The intention was to investigate the urban heat island effect in a populous country other than the U.S., where a significant upward temperature bias due to urbanization had already been demonstrated. If a similar bias were to be found in China, this would add considerable weight to the notion that the heat island effect affects temperature measurements worldwide. If there was little or no difference between Chinese data from urban and rural centers, then the U.S. result could be disregarded as atypical. Much was riding on the two studies.

Both papers included the statement that the weather stations were chosen "... based on station histories: the selected stations have relatively few, if any, changes in instrumentation, location, or observation times." Avoiding the selection of any stations that had moved during the 30-year period of the studies was important, since a move could affect the temperature measurements.

It's this statement that forms the basis of the fraud charges by Keenan, who has accused Wang of fabricating more than half the station histories – that is, making up the temperature data itself – as well as lying about station moves.

The two research papers considered the same 84 meteorological stations in China, these being selected from a larger set of 260 stations identified by a joint project of the U.S. Department of Energy (DOE) and the Chinese Academy of Sciences (CAS) on climate change. Unfortunately for Wang, it turns out that data did not exist for all of these stations.

A DOE/CAS project report issued the year after Wang's paper was published states that, for 49 of the 84 stations supposedly picked out by Wang, "station histories are not currently available" and "details regarding instrumentation, collection methods, changes in station location or observing times ... are not known".[79] For those 49 weather stations, therefore, the statement in the published papers that they were chosen based on station histories cannot possibly be correct.

And for the remaining 35 weather stations that did have data available, even the statement that there were few moves is false as well. An analysis by Keenan reveals that many of these stations moved multiple times during the 30 years of the study; one station had five different locations from 1954 to 1983, some of the locations being as much as 41 kilometers (25 miles) apart.[80] Another station included in the papers was originally in the center of a city and then moved 15 kilometers (9 miles) away to the cooler sea shore. The DOE/CAS report adds:

> Few station records included in the data sets can be considered truly homogeneous [had no changes over the observation period]. Even the best stations were subject to minor relocations or changes in observing times, and many have undoubtedly experienced large increases in urbanization.[79]

According to Keenan, such changes in location could have affected the measured Chinese temperatures that were used in the two studies by as much as 0.4° Celsius (0.7° Fahrenheit),[76] which is a substantial portion of the average temperature increase worldwide from global warming.

Together with the alleged fictitious data used for the other 49 stations, the station moves clearly invalidate Wang's study. A request from Keenan for Wang to retract his findings was ignored, leading to the present fraud charges.

The other study that Wang co-authored, in which the lead author was climate scientist Phil Jones, concluded – partly from the same fabricated Chinese data – that urbanization has no significant effect on measured temperatures.[78] This particular study is important, as it was cited by the IPCC in its most recent report as evidence that the urban heat island effect is negligible,[81] along with its own fabricated evidence discussed in the previous section.

But Jones has recently reversed himself. In a new study of Chinese temperatures, he reported not only that a strong urban warming effect does indeed exist in China, but also that urban warming caused the China-wide temperature rise between 1951 and 2004 to be overstated by a whopping two thirds.[82] This is twice as much as the IPCC's one third exaggeration of global temperatures that I discussed earlier in the chapter. In an obvious attempt to dissociate himself from Wang and redeem his own reputation, Jones did not include Wang among his co-authors on the new study.

We can see to what lengths the IPCC and its followers will go to uphold their position that man-made CO_2 causes global warming. This is the second known case of fabrication concerned with urban heat islands, which means that the temperature bias from urbanization is apparently as big a thorn in the IPCC's side for modern temperature data as were the Medieval Warm Period and the Little Ice Age for the historical record, leading to the infamous hockey stick.

IPCC Dirty Tricks: Politics Ousts Science

The IPCC repeatedly crosses the line between science and politics. As leading climatologist Richard Lindzen of MIT puts it:

> When an issue becomes a vital part of a political agenda, as is the case with climate, then the politically desired position becomes a goal rather than a consequence of scientific research. This paper ... will show ... how scientists adjust both data and even theory to accommodate politically correct positions ...[83]

Nowhere is corruption in the IPCC more visible than in the review process by which the conclusions of its climate scientists become part of the panel's published reports – reports that constitute the bible of global warming alarmists. But just as bias toward human-induced global warming is built into the IPCC's original mandate, politicization and corruption of the review process are, amazingly enough, inherent in the panel's working principles. These principles state:

> Changes (other than grammatical or minor editorial changes) made after acceptance by the Working Group or the Panel shall be those necessary to ensure consistency with the Summary for Policymakers or the Overview Chapter.[84]

In other words, the science should be modified if needed to conform to the Summary and Overview, which get written first! That's like asking a jury to rubber stamp a verdict that the judge has already decided on. In the financial world, it's called cooking the books.

Vincent Gray, who has had firsthand experience with the IPCC report writing process for many years, is a disillusioned IPCC reviewer, though this may be in part because he makes an unusually large number of comments on the reports. Gray says in the beginning, he was convinced that the IPCC would answer queries honestly, and that scientific debate would take place purely on the basis of facts, logic and established scientific principles.[85] But what he found was entirely different:

TABLE 3.2: BAD IPCC SCIENCE – CORRUPTION

Areas of IPCC corruption

- The effect of urbanization on global temperatures: two separate cases.

- The historical temperature record.

- The IPCC's internal process for reviewing reports before publication.

WHAT'S WRONG WITH THIS SCIENCE?

1. The IPCC has dismissed sound statistical evidence that urbanization artificially inflates global temperatures, by using double-talk and by citing nonexistent counter-evidence – which has led to *accusations of fabrication* against the IPCC by the authors of the statistical study.

2. Also connected to the effect of urbanization on temperature, *a formal charge of fraud* has been brought against an IPCC climate scientist, for allegedly fabricating Chinese temperature data and lying about weather station histories. The case is currently the subject of an investigation by the New York Attorney General's Office.

3. To validate the CO_2 global warming hypothesis, the IPCC was complicit in skewing historical temperature data to produce *the deceptive hockey stick curve.*

4. IPCC principles require the science in its reports to conform to the policymakers' summary, a political document that is often written before the body of the report. This is a *corruption of normal peer review.* Selection of IPCC reviewers is also a corrupt process.

Penetrating questions often ended without any answer. Comments on the IPCC drafts were rejected without explanation, and attempts to pursue the matter were frustrated indefinitely. ... the data collection and scientific methods employed are unsound. Resistance to all efforts to try and discuss or rectify these problems has convinced me that normal scientific procedures are not only rejected by the IPCC, but that this practice is endemic, ... I therefore consider that the IPCC is fundamentally corrupt.[86]

A notable example of corruption in the IPCC reviewing process occurred during preparation of the Second Assessment Report back in 1995, when the IPCC was trying very hard to establish the concept of a "discernible human influence" on climate. This has been described in detail elsewhere, so I'll only dwell on it briefly here.

In order to link global warming with man-made CO_2, the IPCC attempts to apply a technique called "fingerprinting", whereby geographic and temporal patterns of higher temperatures (or increased rates of warming) are matched to the predictions of computer climate models.

But the 1995 IPCC claim to have successfully identified a human fingerprint on global climate was false, because it was based on selective use of temperature data – the cherry-picking syndrome once again. In this case, when all the data were examined as a whole, the human fingerprint was found to be absent.

That caused a considerable ruckus at the report review stage. In a rare display of honesty by the IPCC, the draft of the report disputed the success of fingerprinting and actually questioned the evidence for any human effect on climate. The draft included as many as 15 statements reflecting these views, for example: "None of the studies cited above has shown clear evidence that we can attribute the observed [climate] changes to the specific cause of increases in greenhouse gases."[87]

In the final version, this and other similar language had gone, being replaced by quite different statements implicating greenhouse gases such as CO_2 in global warming, and claiming that, for the first time, evidence now existed for a human influence on climate.[88] This claim represented a sea change in the public declarations of the IPCC, which up to that point had made only tentative statements about human-induced global warming, and played a major role in subsequently shifting the majority of public opinion toward the alarmist viewpoint.

But the drastic changes from the draft version of the report had all been made surreptitiously, by a small group of about six of the lead authors for that particular chapter – without the other authors or any reviewers even being consulted. The alarmists had won their first battle against the skeptics, inside the IPCC itself.

The story was exposed in the press around the time the report was published, in June 1996, generating a series of heated exchanges lasting several months in the pages of *The Wall Street Journal* and other publications. Frederick Seitz, an eminent physicist and past president of both the National Academy of Sciences and the American Physical Society, wrote that he had never witnessed "a more disturbing corruption of the peer-review process" than the events that led to the IPCC report, and accused the IPCC of tampering with science for political purposes.[89]

Even the prestigious science journal *Nature*, while supportive of the IPCC's stance on global warming, chastised the organization in an editorial, saying that:

> IPCC officials claim ... the revisions [were made] in particular to ensure that it conformed to a 'policymakers' summary' of the full report ... But there is some evidence that the revision process did result in a subtle shift ... that ... tended to favour arguments that aligned with the report's broad conclusions.[90]

The IPCC, in line with its corrupt working principles, had turned the review process upside down and had changed perfectly good science to match what had already been written in the report summary – instead of following the normal procedure of having the summary conform to the underlying material in the report. One of the authors of the altered chapter basically admitted as much in a later attempt to defend the IPCC's position, saying that the review procedure was indeed required by IPCC rules, and that the whole report was unquestionably "fraught with political significance".[91]

Little has changed at the IPCC since then. The corruption permeates not only the report review process, but also the selection of authors and reviewers. It goes without saying that the IPCC generally selects authors who agree with its position on human-induced global warming.

Review of the recent Fourth Assessment Report's physical science section involved more than 650 individuals, as well as governments and international organizations.[92] But of the 308 reviewers of the second draft, about a third were also authors of at least one chapter of the report,[93] an unethical practice that creates conflict of interest and is open to abuse in itself. The peer review process in conventional scientific publishing never allows authors to review and promulgate their own work.

The second draft of the pivotal Chapter 9, which deals with the attribution of global warming to human activity, was examined by 68 reviewers.[94] Out of these, some again were contributing authors of the very same chapter or other chapters of the report. Others were government reviewers with a vested interest in the chapter's conclusions, as their governments have policies on limiting CO_2 emissions and therefore implicitly accept the IPCC's fundamental claim about human-induced warming.

The small number of reviewers of Chapter 9 who could be considered truly independent were far outnumbered by those partial to the IPCC belief. And even though the IPCC's working principles allow for the inclusion in the final report of "different (possibly controversial) scientific, technical, and socioeconomic views on a subject, particularly if they are relevant to the policy debate"[95] – which in theory could accommodate opposing opinions from reviewers – this rarely happens in practice.

So it's really not surprising that the chapter's conclusions endorse the CO_2 global warming hypothesis and ignore all conflicting evidence.

Incidentally, both global warming skeptics *and* alarmists have alleged that the report production process is corrupted. To the skeptics, the IPCC review process is so corrupt that they can hardly get the organization's attention. The alarmists, who as we've just seen dominate both the writing and reviewing of IPCC reports, have complained that the reports are often watered down at the review stage in order to appease skeptics. Either way, the reviewing process is a far cry from good science.

Chapter 4: Computer Snake Oil

It's no secret, but it's not widely known that the gloom and doom about CO_2 and global warming preached by the IPCC is founded entirely on computer models. These models can be very powerful tools, but they can also be wrong.

One area where computational models are particularly useful is the design and engineering of complex technological marvels such as cars, airplanes, or computer chips. For all of these, the underlying science is well-known and the assumptions behind the models have been thoroughly tested. Computer modeling also works well in some of the social sciences such as population forecasting, for the same reasons.

The models don't do as well when the computer calculation is unable to mirror the reality of a complex system. This happens when there are gaps in our knowledge of the model's infrastructure, and we have to use guesswork to fill in the gaps. Now we can no longer be certain about what the model tells us. As a book on the subject, *The Promise And Limits Of Computer Modeling*, puts it:

> Is [a computer model] an image of the world? Or is it a glorified fiction, like a motion picture or video game? ... the computer model inspires not complete confidence, but an equivocation, a murkiness about what the model really means.[96]

How good are the predictions made by computer climate models? Because this question is central to the global warming debate, the whole of this chapter is devoted to it.

IT'S ONLY A MODEL

Computer simulations are only as reliable as the assumptions that the computer model is built on: "garbage in, garbage out", as software engineers like to say. Just as scientific hypotheses need to be tested experimentally before they can be confirmed – a requirement that the IPCC recognizes[97] – so do any educated guesses made in a computer model.

One of the biggest defects in the IPCC's argument for human-caused global warming is that its climate models are chock-full of untested assumptions. They include conjectures about basic elements of the climate system, such as clouds and precipitation; about the contribution of natural cycles to current high temperatures; and about the physical processes that control the response of the Earth's climate to tiny additions of CO_2.

We'll discuss these and other assumptions as we go through the book. The real problem is that many of the assumptions are undergoing testing right now, in an ongoing experiment that may not be complete for another 50 or 100 years. Alarmists and the IPCC say we can't afford to wait that long for the answers, so our only choice is to rely on computer modeling.

Nevertheless, the evidence is mounting that the IPCC's computer models may not accurately simulate even the present-day climate, let alone the future or the past. Many of the predictions made by these models about current climatic conditions have been dead-wrong, and the only reason that climate models can "hindcast" the historical record is that they are arbitrarily adjusted to fit the historical data.

Even though computer simulations can be useful, it's easy to become carried away with their possibilities and to inflate the importance of their predictions.

I fell into this trap myself at the beginning of my career, when computer simulations were all the rage following the development of powerful mainframe computers in the 1960s. I was doing research that involved bombarding solid materials with high-speed atoms from a small accelerator, as a means of probing the material structure – a new research field at the time.

The experiments were complex and interpretation of the results was sometimes unclear. So when one of my colleagues introduced to me to computer modeling, I jumped at the opportunity to try it out. That way, I could simulate the whole experiment on a computer and, by changing the variables at will, gain some insight into the accelerator data. Maybe, I thought, the approach would be so successful that I could even give up the accelerator experiments, which were tedious and often lasted for 24 hours or longer at a stretch.

However, this hope was quickly dashed. The computer model was fun to learn, and I was soon churning out predictions for my experiments by the dozen. The only trouble was that the computer simulations didn't match the real world too well!

While some features of the accelerator results were reproduced by the computational model, other features weren't, and the computer calculations tended to exaggerate the magnitude of trends in the experimental data. The reason was simple: the computer model was based on assumptions about the bombarding atoms and the target material, and not all of those assumptions were correct. Refining the assumptions, over a long period of time, led to better agreement between simulation and experiment, but never to the point where I could rely on the computer model for accurate prediction. There were just too many unknowns that couldn't be measured or tested.

It's not very different with climate models today. Some of the models are very elaborate and can predict all sorts of climatic variables in great detail – so much so that we can be awed by the model itself, by the sheer calculating power at our fingertips. I've certainly experienced that feeling.

But a model is just a model. Although the computing power of current supercomputers used for climate modeling is more than a million times greater than what was available in 1970, that doesn't change the fact that the models are constrained by their underlying assumptions. All the computing speed in the world doesn't make up for lack of understanding.

The 2008 Wall Street meltdown is a painful reminder of this. Although the reasons for the U.S. financial crash and the continuing global recession are numerous, there's no doubt that computer models played a role, since it was computer simulations that had been routinely used to evaluate risk by the investment community. If the models had better represented reality, or if bankers had been more aware of the limitations of computer modeling, perhaps the financial crisis wouldn't have been as bad as it is.

Another notorious example, from the engineering field, is the Millennium Bridge in London. It was only after the footbridge was built and people walked on it for the first time that unexpected swaying was felt, causing the bridge to be promptly closed for design modifications. The engineers realized they had created what is known as a resonant structure, a well-understood phenomenon that could have been avoided if they had made the right assumptions in their computer models of the bridge.

Climatologists defend their computer climate models by saying the models are the only handle we have on the climate. That is true, unless we're prepared to wait several decades until we have some more definitive measurements. But that doesn't mean the models are right. I certainly wouldn't want to fly on an airplane designed by the engineering equivalent of a computer climate model.

Remember, it's only a model.

FITTING AN ELEPHANT

Ideally, what we'd like in a climate model is an accurate depiction of the most important climatic features – especially those important to humans, such as temperature, precipitation, winds, and storms.

That's easier said than done, as the model must not only obey the laws of physics and chemistry, but it must also simulate a host of complex interactions in the Earth's climate system, which couples the atmosphere to the land masses to the oceans to snow and ice.[98] This complexity includes phenomena as diverse as jet streams in the upper atmosphere, deep ocean currents, clouds, greenhouse gases, and climate cycles such as El Niño.

Climate processes span enormous ranges of distance and time, from a few centimeters to thousands of kilometers, and from several hours to millennia. For a computer simulation, all the processes and their interactions must be expressed in the form of mathematical equations, which are translated into computer codes. Armed with these equations, the computer then simulates how the climate evolves with time.

That may sound straightforward, but there are two big limitations. The first is that even the most powerful computer in the world today is not capable of simulating the climate on a small spatial scale. So a grid is employed, with the grid boxes for atmospheric models being about 200 kilometers (120 miles) square across the Earth's surface, and about 0.75 kilometers (0.5 miles) high vertically; ocean grids are finer.

This means that anything smaller, including cloud formation, falls between the cracks of the grid and can't be modeled accurately. For small-scale processes, simplified pictures of reality involving approximations become necessary. Some approximations aren't as good as others.

The second major drawback to computer modeling is that, for all our modern technological prowess, there's plenty we don't understand or don't even know about the large-scale workings of the Earth's climate over large distances. That's where the assumptions come in. And aside from the fact that many assumptions can't be tested, as I've said before, the assumptions require yet more approximations.

All these approximations, small-scale and large-scale, are incorporated in the model in the form of adjustable numerical parameters.[99]

The very existence in computer climate models of so many adjustable parameters, often termed "fudge factors" by scientists and engineers, should be a warning sign in itself. The famous mathematician John von Neumann once said, "With four [adjustable] parameters I can fit an elephant, and with five I can make him wiggle his trunk."[100]

Some adjustable parameters simply cover a range of error in a measured quantity, but others are unknown and have to be guessed. To adequately describe fine-scale processes that take place inside one of the 200-kilometer-square grid boxes over the Earth's surface, almost 30 parameters alone are needed.[101] With many more required to account for all the large-scale assumptions, the total number of parameters in a climate model can run into the hundreds. That's a lot of elephants!

Climate scientists have come up with several procedures to get a grasp of some of the unknowns in their models. One of these is to use more than one model, each based on slightly different variable parameters, in order to estimate the uncertainty in the parameters. Another method "tunes" selected parameters until the model correctly represents known quantities, such as the global energy balance.[102]

But remember, it's only a model. No fixes to present-day computer models can escape the fact that the models still depend on numerous untested assumptions and adjustable parameters. Even the very latest models contain biases arising from incorrect assumptions, as pointed out in a recent report to Congress by U.S. climate modelers:

> Nonetheless, there are still systematic biases in ocean-atmosphere fluxes in coastal regions west of continents, the spectrum of ENSO [El Niño–Southern Oscillation] variability, spatial distribution of precipitation in tropical oceans, and continental precipitation and surface air temperatures.[103]

With so many parameters uncertain or unknown, the predictions of any of these models become dubious, to say the least. Yet the IPCC continues to say it's 90% certain of the link between global warming and man-made CO_2 – a prediction based solely on computer climate models. Tweak the parameters differently, and the models will predict little or no warming.

And that's just the present climate (Table 4.1). The IPCC constantly warns that global warming will get worse and, believe it or not, the panel has actually forecast how hot it will be 90 years from now, using the very same computer models. Talk about stretching the truth.

The IPCC forecasting process[104] has recently been audited by Scott Armstrong, a marketing professor who is an internationally recognized expert on forecasting methods in general, and a colleague.[105] The audit included a survey of IPCC authors and reviewers, together with a smaller number of known global warming skeptics. It found that not only did the IPCC's procedures violate a majority of standard forecasting principles, but also:

> The forecasts in the Report were not the outcome of scientific procedures. In effect, they were the opinions of scientists transformed by mathematics and obscured by complex writing. ... Extensive research has shown that the ability of models to fit historical data has little relationship to forecast accuracy.[106]

So much for the IPCC's climate forecasts. Remember, it's only a model.

CLOUDING THE PICTURE

In computer climate models, it is assumptions about two watery entities – clouds and atmospheric water vapor – that underpin the IPCC's conclusions on man-made global warming. According to the models, the normally tiny effect of CO_2 on global temperatures is amplified most by water vapor and clouds.

The two are related, since clouds are born when water evaporates to form water vapor, which later condenses into liquid droplets or ice crystals to produce the clouds. But they are not the same, clouds being essentially liquid water while water vapor in the atmosphere is a greenhouse gas, so they affect temperature and climate differently. We'll come back to water vapor in the next chapter, which deals with the sensitivity of the Earth's climate system to CO_2.

The problem with clouds is that they can't be properly modeled in present computer simulations. This is partly because we simply don't know much about what goes on inside a cloud either during the cloud's formation, or when it rains, or when the cloud is absorbing or radiating heat. The result is a lot of adjustable parameters.[107]

The inability to model clouds is also partly because actual clouds are much smaller than the computer grid scale, by as much as several hundred or even a thousand times. Even the coming generation of supercomputers will be able to represent only the largest clouds accurately.[108] So clouds are represented in computer models statistically – that is, by average values of size, altitude, number, and geographic location. You know that's not a very good representation just from watching the nightly weather forecast.

Inadequacies in computer simulations of clouds are acknowledged by climate modelers, even though these same modelers insist that the models can be used to make highly reliable predictions about the future. James Hansen, who heads NASA's Goddard Institute for Space Studies and has carried out computer climate simulations for the last 30 years, admits in a recent research paper that:

> Model shortcomings include ~25% regional deficiency of summer stratus cloud cover off the west coast of the continents with resulting excessive absorption of solar radiation by as much as 50 W/m², deficiency in absorbed solar radiation and net radiation over other tropical regions by typically 20 W/m², sea level pressure too high by 4–8 hPa in the winter in the Arctic and 2–4 hPa too low in all seasons in the tropics, ~20% deficiency of rainfall over the Amazon basin, ~25% deficiency in summer cloud cover in the western United

States and central Asia with a corresponding ~5° Celsius excessive summer warmth in these regions.[109]

Apart from all the other deficiencies listed, especially for clouds, it's worth noting that this particular model's summer temperature overestimate of 5° Celsius (9° Fahrenheit) in the U.S. and Asia is as large as the drop in global temperatures that accompanied the last ice age! With such big errors, how can anyone expect models like this to accurately simulate global warming of less than 1° Celsius (1.8° Fahrenheit) to date?

Hansen, who is also one of the shrillest global warming alarmists, talks in the same paper about "dangerous anthropogenic interference" with the Earth's climate caused by CO_2, a phrase that comes from the UN.[110] It's probably Hansen and his climate model that are the danger to humanity.

Interestingly enough, the IPCC seems aware of the limitations of its computer models for clouds. Chapter 8 of its 2007 report states:

> Nevertheless, models still show significant errors. ... many important small-scale processes cannot be represented explicitly in models, and so must be included in approximate form ... Significant uncertainties, in particular, are associated with the representation of clouds, and in the resulting cloud responses to climate change.[111]

Chapter 8 goes on to say that different computer climate models vary considerably in their estimates of sensitivity to CO_2, mainly because of differences among the models in the way that the warming effect of CO_2 is amplified by clouds.[112] And then Chapter 10 reveals that just the uncertainty alone in the predicted heating and cooling (radiative) effect of clouds is larger than the total modeled warming anticipated from a doubling of CO_2 over its preindustrial level.[113]

Yet, despite this recognition that its computer simulations are deficient in modeling clouds – which contribute almost as much as water vapor to CO_2 amplification in the models – Chapter 9 of the very same IPCC report claims that global warming over the last 50 years "very likely" comes from CO_2.[114] This is part of an all-too-common syndrome that permeates the IPCC's reports, where the left hand doesn't appear to know what the right hand is doing.

As we'll see shortly, clouds may hold the key to explaining global warming, but for reasons entirely unrelated to CO_2 and human activity.

TABLE 4.1: COMPUTER CLIMATE MODELS

What they include

Atmosphere	– Surface air temperature, pressure and humidity
	– Heat from the sun
	– Clouds and precipitation/evaporation
	– Winds
	– Aerosols
	– Natural and man-made greenhouse gases
Land surface	– Soil temperature and moisture
	– Vegetation and biological processes
	– Rivers and lakes
	– Snow cover
Oceans	– Ocean temperature and salinity (saltiness)
	– Atmosphere-ocean coupling
	– Deep currents
Sea ice	– Reflection of sunlight
	– Ice movements
	– Melt ponds.

WHAT'S WRONG WITH THESE MODELS?

1. *It's only a model.* Computer simulations are only as accurate as the underlying assumptions. Climate models are full of untested assumptions, which are embodied in the models as adjustable "fudge factors".

2. *Clouds.* The formation and behavior of clouds are very poorly understood, so cannot be modeled with any confidence. Yet cloud "fudge factors" are crucial to estimating the response of the climate to CO_2.

3. *Oceans.* The models predict warming of both the atmosphere and oceans, but the oceans have stopped heating up since 2003.

4. *Precipitation.* Deficiencies exist in the computer modeling of both tropical rainfall and extremely heavy rainfall.

5. *Cyclic events.* Several short-term climate cycles such as El Niño are not well simulated by present computer models.[115]

6. *Indirect solar effects.* Indirect effects from the sun, such as shielding of cosmic rays that create cooling clouds, are not included.

FAILED PREDICTIONS

The ultimate test of any computer model is how accurately it represents reality. To evaluate its computer climate models, and to "validate" them for predicting the future, the IPCC uses the models to simulate past and present climates.

The difficulty with this approach is that many climate models can only make predictions about the present (and future) if they are first matched to the past climate record, because of the need to pin down some of the hundreds of variable parameters that I discussed before. But the uncertainty in the parameters makes all predictions unreliable at best and, for much of the present-day climate record, just plain wrong.

Since we're talking about global warming, you'd expect the computer models to correctly predict temperatures if nothing else. However, the models' track record is about as impressive as a football team that fails to score for the whole season.

The Arctic and Antarctic

One of the predictions of computer climate models is that CO_2-driven global warming should be strongest at the North and South Poles. This is partly because of a chain reaction in which melting snow and ice expose darker surfaces underneath that soak up extra sunlight, causing further melting.[116]

In the Arctic, although temperatures are currently rising quite rapidly, there has been no *net* warming since 1937[117] – in contrast to global temperatures, which have gone up perhaps 0.4° Celsius (0.7° Fahrenheit) since then (see Figure 2.1). There has been an overall upward trend in the Arctic since 1900 similar to that seen across the whole Northern Hemisphere, but the strong, multidecadal Arctic temperature swings are far from anything predicted by IPCC computer models.

Global warming alarmists have made much of shrinking Arctic sea ice, supposedly caused by global warming. However, the shrinkage has recently reversed – a topic we'll return to later – and, in any case, was most likely due to a natural climate cycle rather than higher CO_2 levels.[118]

In the Antarctic, the picture is more complicated. While West Antarctica and the small Antarctic Peninsula which points toward Argentina are both warming, the remaining 80% of the continent has shown no significant temperature trend since the 1960s, and may even be cooling.[119]

A research paper published in early 2009 created a buzz by claiming that the whole Antarctic continent, and not only the Western portion, has been warming for the last 50 years.[120] The claim was based on a statistical analysis, in which temperatures across the continent were reconstructed from the sparse temperature record of Antarctic weather stations, most of which are near the coast. As just mentioned, the actual record shows steady temperatures for the continent as a whole, unlike computer climate models that predict universal warming.

But just as the hockey stick was the result of misapplication of statistical techniques, the reconstructed Antarctic temperature record founders on its statistical analysis. Economist Hu McCulloch has shown that when the reconstruction is carried out correctly, the warming trend for West Antarctica and the Peninsula still stands, but "the trends for the continent as a whole and for East Antarctica are not significantly different from zero".[121] You probably won't be surprised that the authors of the dubious research paper include Michael Mann of hockey stick infamy,

And, despite claims to the contrary, the sea ice around Antarctica has been expanding recently.[122]

Northern vs Southern Hemisphere

A second temperature prediction that computer models don't get right is that the Southern Hemisphere should be warming more than the Northern Hemisphere. That's because of the cooling effect of aerosols, which are tiny particles in the atmosphere, mostly resulting (like CO_2) from the burning of fossil fuels. Aerosols are emitted predominantly in the more industrial Northern Hemisphere, which should therefore get less hot than the Southern Hemisphere, according to the models.[123] But exactly the opposite is true and temperatures are higher in the Northern Hemisphere, as the IPCC documents in its latest report.[124]

Atmospheric Warming

Computer climate models based on the CO_2 theory of global warming predict that the atmosphere, especially the lower atmosphere,[125] should heat up faster than the Earth's surface. But satellite data show the reverse. At altitudes up to at least 10 kilometers (6 miles) in the tropics, which is where the difference is most conspicuous, the atmospheric warming rate is lower than at the surface. And not only are the computer models wrong about that, but they also predict a warming rate up to three times larger than it actually is.[126]

The Oceans

As a fourth example of faulty temperature prediction by computer models, the world's oceans have stopped warming since 2003[127] – even though the models say the oceans should be getting warmer along with land surfaces and the atmosphere. Another shortcoming that the IPCC itself admits to is bias in modeled sea surface temperatures, which are too low in some parts of the Northern Hemisphere, and too high on the eastern side of tropical oceans.[128]

Further deficiencies in the IPCC's climate models include the modeling of rainfall, particularly in the tropics, the simulation of several short-term climate cycles such as the Pacific Decadal Oscillation, and the inability to explain why the amount of the greenhouse gas methane in the atmosphere has suddenly stopped rising.

What a litany of failings! Yet the IPCC relies on these very same models to uphold its assertion that CO_2 emissions are responsible for global warming.

Remember, it's only a model. Until we improve our present rudimentary understanding of many features of the Earth's climate, the models will continue to fall short. Computer climate models are like a boat riddled with holes, with the IPCC and its alarmist disciples frantically bailing to keep it afloat.

Chapter 5: CO$_2$ Sense and Sensitivity

The key to the IPCC's conviction that global warming comes from human CO$_2$ emissions lies in the assumptions behind its computer models – especially those assumptions that influence the so-called sensitivity of the Earth's climate system to CO$_2$. By sensitivity is meant the boost in global temperatures for a particular amount of CO$_2$ added to the atmosphere.

As we saw in Chapter 3, neither temperatures nor the CO$_2$ level may have increased as much as we think. But, regardless of the actual size of the gains, there's no doubt that both temperature and CO$_2$ have gone up in the past 160 years or so, and that at least the CO$_2$ level is continuing to rise right now.

How strong is the temperature - CO$_2$ connection?

From the proclamations of the IPCC and its alarmist supporters, you might think that the whole CO$_2$ global warming issue is cut-and-dried, that there's no question about climate sensitivity, and that there's already enough CO$_2$ in the atmosphere to cause devastatingly high temperatures in the years to come. Nothing could be further from the truth.

Figuring out climate sensitivity from computer climate models relies on a number of questionable, untested assumptions, mostly about the Earth's past climate. So it should come as no surprise that the predictions these models make about climate sensitivity are just as unreliable as the models' faulty temperature predictions that I discussed in the previous chapter.

CO$_2$ FEEDBACK: FALSE POSITIVE?

Climate sensitivity is intricately linked to the important concept of feedback, a technical term borrowed from the field of electronic engineering. It's not a difficult concept to grasp though, as the word has much the same meaning as in everyday life.

In science, feedbacks result in sustained magnification (called positive feedback) or sustained diminution (negative feedback) of a response to a disturbance of the status quo. A weather-related example of positive feedback is the gaining of strength by tornadoes or hurricanes through amplification processes that cause a self-reinforcing chain reaction.[129] The feedback often continues to intensify the storm until some other factor intervenes, such as a hurricane making landfall.

Negative feedback, on the other hand, results in the damping down of a process, so as to bring a system back to where it was before. Negative feedback processes are very common in nature, acting as a safety valve to keep delicate ecosystems and the animal kingdom in balance.

The human body is a remarkable example of many types of negative feedback. Body temperature, blood pressure, blood sugar level and numerous other functions are all controlled by negative feedback mechanisms, which maintain that function within the narrow range necessary for our survival. For example, body temperature is regulated through feedbacks that cause the body to sweat when it's too hot or to shiver when it's too cold, in order to return the temperature to normal.

So why are feedbacks important for global warming? The reason is that without feedback, in fact without net *positive* feedback to enhance the climate sensitivity to CO$_2$, there wouldn't be anything for the alarmists to worry about. Positive feedbacks ratchet up the warming, but negative feedbacks turn it down.

On its own, the 387 parts per million of CO$_2$ in the atmosphere today is not enough to cause even a 0.8° Celsius (1.4° Fahrenheit) rise in temperature since 1850, if the only explanation is the greenhouse effect that I discussed back in Chapter 2. Man-made global warming, if it exists at all, would be insignificant without positive CO$_2$ feedback to magnify the Earth's natural greenhouse effect. Climate modelers say this feedback comes primarily from water vapor and from clouds, with a small contribution from snow and ice.[130]

But if there's no feedback, or if the net feedback is *negative* – a distinct possibility, as we'll see shortly – then global warming is not likely to have been caused by CO$_2$. With negative CO$_2$ feedback, even a future doubling of CO$_2$ from its preindustrial level will have little influence on temperatures.

That's not the story you hear from alarmists and the IPCC, who sermonize that the world is already close to disaster and that we need to rein in our CO$_2$ output immediately. Few people realize that this gloomy prognostication is founded on shaky theoretical computer models that predict positive CO$_2$ feedback.

Most feedbacks in nature are negative, presumably for good reason – to maintain stability in the natural world. Why should the Earth's climate system be any different, even if the perturbation comes from human activity?

Positive feedback, which amplifies the initial disturbance, can lead to runaway conditions and a system out of control. However, the positive feedback mechanism can be held in check if other processes, some of which involve self-correcting negative feedback, are operating at the same time. The fact that no runaway climate events have occurred in the Earth's past suggests again that climate is governed by negative, rather than positive, feedbacks.

The IPCC concludes from its computer climate models that the major global warming feedbacks are all positive, with just one exception (Table 5.1).[131] But there's very little observational evidence to justify this conclusion.

For example, the argument is often made by global warming alarmists that water vapor feedback must be positive, or there would be no way to explain the observed warming. However, this argument is based entirely on computer models that don't allow for any natural sources of global warming other than the sun, the influence of which is underestimated in any case (Chapter 6). If natural sources were included in the models, it's quite possible the models would predict that water vapor feedback is negative.

There's considerable uncertainty too about other feedbacks that can affect global temperatures, such as the carbon cycle feedback, which can pump up the amount of CO$_2$ retained in the atmosphere.[132] Even the IPCC admits these feedbacks are poorly understood:

> Large differences between models, however, make the quantitative estimate of this [carbon cycle] feedback uncertain. Other feedbacks (involving, for example, atmospheric chem-

ical and aerosol processes) are even less well-understood. Their magnitude and even their sign remain uncertain.[133]

Table 5.1: Major CO_2 Feedbacks in IPCC Models

Feedback	Positive or negative
Water vapor[134]	Positive
Clouds	Positive
Temperature/altitude[135]	Negative
Snow and ice	Positive

With few exceptions, we just don't know which climate feedbacks make the most difference nor which ones the least, nor even whether those feedbacks that appear to be positive and destabilizing really are – and not negative and climate-stabilizing instead. Furthermore, many models ignore the fact that different feedbacks are often coupled to one another. The feedbacks deduced from IPCC climate models are no more accurate or reliable than the many adjustable parameters in the models.

Negative Feedback in Satellite Data

Evidence for global warming feedbacks, either positive or negative, is hard to come by. So two recent studies by researchers at the University of Alabama in Huntsville that show strongly negative cloud feedback, based on satellite observations of the Earth,[136] are quite astounding.

The first study probed day-to-day changes in climate variables such as cloud cover, rainfall, and temperature, over a two-month period, for a short-term climate cycle in the tropics.[137] The researchers were surprised to find a decrease in high-altitude cloud coverage as the tropical air warms during the cycle, in direct contradiction to IPCC climate models that predict an increase in high-level clouds from such warming.

A reduction in high-level clouds, which are the clouds that cause heating of the Earth's surface (low clouds cool), means negative feedback is operating – the climate system responds to the natural tropical warming by cooling everything down, trying to return the climate to its previous state.

As lead author and climatologist Roy Spencer explains,[138] the tropical warming cycle serves as a proxy for global warming caused by man-made greenhouse gases such as CO_2. In other words, global warming feedback from high clouds is negative and damping, the very opposite of what the IPCC concludes, which is that CO_2 cloud feedbacks are positive and amplifying. Spencer adds that the satellite data supports an earlier proposal by MIT's Richard Lindzen and a NASA research group that high-level clouds near the equator open up, like the iris of an eye, to release extra heat when the temperature rises[139] – also a negative feedback effect.

The University of Alabama study was criticized, however, on the grounds that it applied only to tropical regions, and that feedback effects identified on such a short time scale (weeks) may not occur on the longer time scales associated with global warming. So Spencer and his team undertook a second study, involving analysis of data from another satellite,[140] which also provides solid evidence for negative cloud feedback.[141]

The second study compared cloud feedbacks deduced from five years of satellite data, collected over the global oceans, with the same feedbacks calculated from IPCC climate models. Not only did the new study reveal the same distinctly negative cloud feedback[142] as the original satellite study, but the comparison also showed that none of the IPCC models displayed the negative feedback behavior seen in the satellite data. In fact, the cloud feedbacks from all the climate models were positive, just as the IPCC insists they are (see Table 5.1).

In contrast to the first study, which demonstrated negative feedback from a reduction in high-level clouds that warm the Earth, Spencer says the second study appears to show negative feedback from an increase in low-level clouds, which exert a cooling effect. But both studies prove that the IPCC's conclusions about cloud feedback[143] are wrong, over at least a five-year time frame.[144]

If cloud feedback is negative rather than positive, and as strongly negative as the University of Alabama studies indicate, then it's entirely possible that combined negative feedbacks in the Earth's climate system dominate the positive feedbacks from water vapor, and from snow and ice (Table 5.1).[145] This would mean that the overall response of the climate to added CO_2 in the atmosphere is to diminish, rather than magnify, the temperature increase from CO_2 acting alone – the reverse of what the IPCC claims is happening. Of course, it also means that global warming must have some other explanation.

Before we examine alternative explanations for global warming, we'll look at how the IPCC's predilection for positive CO_2 feedback leads to overestimation of the effect that human CO_2 emissions have on temperature.

CO_2 OVERSENSITIVITY

To come up with specific numbers for climate response to CO_2, the IPCC and its computer modelers generally go back to the past and use their models to reproduce the historical climate record.[146] This record embraces data for the last 160 years, during which the atmospheric CO_2 level has climbed significantly; proxy data for the last millennium; and ice-core proxies for the long-ago ice ages (the paleoclimate).

The models include adjustable parameters that affect the various feedbacks just discussed, as well as parameters to describe what are called forcings – the actual disturbances that alter climate and that give rise to feedback, such as radiation from the sun, greenhouse gases in the atmosphere, and aerosols. Forcings can be positive or negative, depending on whether they produce a heating or cooling effect, respectively.

Just what are climate sensitivity numbers? Climate sensitivity is usually expressed as the temperature increase caused by a doubling of the atmospheric CO_2 level over its preindustrial level in 1850, which is about when the present period of global warming began (Figure 2.1).

With high climate sensitivity, which means a climate exceptionally sensitive to extra CO_2, the temperature increase for doubled CO_2 will be large – as much as 4.5° Celsius (8.1° Fahrenheit) or even more, according to some IPCC models.[147] On the other hand, with low climate sensitivity, the temperature rise for twice the CO_2 will be a lot smaller – perhaps 2° Celsius (3.6° Fahrenheit) in the IPCC's view.[148] At its present rate of increase, the CO_2 concentration in the atmosphere will double sometime around the year 2100.

Table 5.2 shows the temperature gain since 1850 at doubled CO_2 (our definition of climate sensitivity), as well as at today's CO_2 level, as estimated by various climate models. Information about the models can be found in Appendix A, together with details of the calculations. Because the CO$_2$ concentration hasn't doubled yet, the predicted temperature increases for today's climate are modest.

Table 5.2: CO$_2$ Climate Sensitivity (degrees Celsius)[149]

Model	Predicted temperature increase from CO$_2$	
	Today	At doubled CO$_2$
IPCC (2001)		3.5 °C
IPCC (2007)	0.76°C	3.3 °C
Hansen	0.6 °C	2.7 °C
Spencer	0.21 °C	0.45 °C
Zero CO$_2$ feedback	0.54 °C	1.2 °C
Positive CO$_2$ feedback	above 0.54 °C	above 1.2 °C
Negative CO$_2$ feedback	below 0.54 °C	below 1.2 °C

Also shown in Table 5.2 are the anticipated temperature increases for different types of feedback in the Earth's climate system: zero (no net feedback), positive, and negative.

You will see immediately why feedbacks are central to climate sensitivity (the CO$_2$ amplification problem in Chapter 2). If the net CO$_2$ feedback is positive, the temperature at doubled CO$_2$ in 2100 could well be as high as the IPCC is saying, which is several degrees Celsius above its 1850 value. But if the net CO$_2$ feedback is negative, the increase from 1850 to 2100 will be no more than about 1° Celsius (1.8° Fahrenheit), possibly even less – in which case we don't have much further to go, considering that the thermometer has already risen 0.6° Celsius (1.1° Fahrenheit).[150]

You can also see from Table 5.2 that the sign (positive or negative) of the net CO$_2$ feedback dictates how much of today's global warming originates from CO$_2$, and how much comes from other sources. If the net feedback is positive, as in IPCC climate models, almost all of the 0.6° Celsius temperature rise can be ascribed to CO$_2$. However, if the net feedback is negative, CO$_2$ can account for only a small portion of the temperature increase to date,[151] and the rest must have other origins.

As discussed in the previous section on satellite data, there is now evidence that cloud feedbacks are negative, instead of positive as concluded by the IPCC. Negative cloud feedback that is sufficiently strong to overcome any positive feedbacks in the climate system will make the overall feedback negative – and make the CO_2 climate sensitivity too small to be of any concern.

Negative feedback implies low climate sensitivity, which means small temperature increases, even for lots more CO_2 in the atmosphere. Or no global warming from man-made CO_2. It's as simple as that.

Feedbacks aside, there are still several issues with the historical approach to validating computer climate models. By far the biggest issue, and one of the basic weaknesses of the IPCC's models, is the underlying assumption that the Earth's climate sensitivity has remained unchanged throughout history, forever the same.

Past climate conditions were often very unlike today's, with temperature swings up to 10° Celsius (18° Fahrenheit) as the planet switched between frigid ice ages and warmer interglacial periods; temperatures that varied more slowly with time than now; different levels of CO_2 and other greenhouse gases; vast ice sheets that built up and then disappeared thousands of years later; different types and amounts of vegetation; and more.

Why should the climate sensitivity have stayed the same while the Earth's climate changed so much?

It's indeed possible that climate sensitivity is some sort of universal constant, like the speed of light, but it's also possible that it varies along with the climate. There's no particular reason that it shouldn't. The IPCC admits as much, even though all its climate predictions are based on the assumption of unaltered climate sensitivity through the ages:

> The use of a single value for the ECS [equilibrium climate sensitivity] further assumes that it is constant in time. However, some authors ... have shown that ECS varies in time in the climates simulated by their models. Since results from instrumental data and the last millennium are dominated primarily by decadal-to centennial-scale changes, they will therefore only represent climate sensitivity at an equilibrium that is not too far from the present climate.[152]

Just how difficult it is to fit the historical climate record, we've already seen in the hockey stick fiasco. So intent was the IPCC on making temperatures for the last 2,000 years conform with the CO_2 record (the CO_2 steady level problem) that it ignored historical temperature data showing otherwise, and indulged in deceptive data manipulation.

Climate modelers may not have sunk to the same depths, but they can't escape the fact that fitting computer simulations to historical data does not yield a unique number for climate sensitivity. This is because there are so many variable parameters in the models. The only way the climate sensitivity can be deduced at all is by making assumptions about all the other adjustable parameters – assumptions that may not be correct. Remember, you can fit an elephant with only five parameters.

We'll look now at the difficulties that arise in trying to use computer models to simulate the climate record from the ice ages (the CO$_2$ lag problem).

TWO-FACED CO$_2$

As I've already said, climate sensitivity numbers generally come from hindcasting, from matching the output of computer climate models to historical data. Some of that data we've encountered in previous chapters. Figure 5.1 presents data showing the temperature record and CO$_2$ level in the Earth's atmosphere over the past 150,000 years, including the most recent ice age that lasted about 100,000 years and ended 11,000 years ago.

Figure 5.1: The CO$_2$ – Temperature Lag

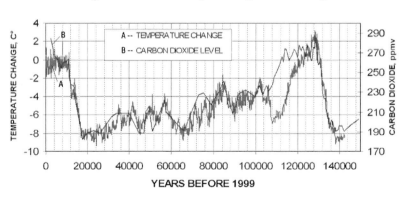

Source: Pangburn.[153] *Note that time in this chart advances from right to left, with the present on the extreme left of the figure and the distant past on the right.*

Something to notice about this data, which is obtained from analysis of Antarctic ice cores,[154] is that the CO$_2$ level closely mimics changes in temperature, but the CO_2 lags behind – with CO_2 concentration changing, up or down, well after the corresponding temperature shift occurred. The lag is about 600 to 800 years,[155, 156] and may have been even longer hundreds of thousands of years ago.

Most paleoclimatologists believe that CO_2 lagged temperature during the ice ages because it takes several hundred years for CO_2 to come out of, or get into, the world's oceans, which is where the bulk of the CO_2 on our planet is stored. The oceans can hold much more CO_2 (and heat) than the atmosphere. Warm water holds less CO_2 than cooler water, so the oceans release CO_2 when the temperature goes up, but take it in as the Earth cools down. The lag time is related to what oceanographers call the ocean-mixing time for CO_2.[154, 157]

Many skeptics quickly jumped on the lag as evidence against the CO_2 global warming hypothesis, which requires warming from human CO_2 to follow the rising CO_2 level, not the other way around. The hypothesis appears to be incompatible with lagging CO_2 under any circumstances.

What those skeptics failed to appreciate is that CO_2 trailed temperature in past global warming at the end of ice ages because of the 600 to 800-year delay time for oceans to give up stored CO_2. In the present warming stretch that has spanned 160 years so far, none of the CO_2 lurking in the oceans has appeared yet, and presumably won't for about another 450 to 650 years, due to the lag. But the CO_2 that we're producing ourselves is completely mixed into the atmosphere within a year or two, and therefore available almost immediately for greenhouse warming and feedback. So it's at least *possible* for temperature to closely follow CO_2 today, though not in the past.

However, it turns out that the CO_2 lag still deals a deathblow to the notion of man-made global warming, for a different reason. The reason has to do with the behavior of the temperature and CO_2 level at the cessation of an ice age.

Ice ages are believed to have ended (and begun) because of changes in the Earth's orbit around the sun.[158] After tens of thousands of years of bitter cold, the temperature suddenly took an upward turn. As we've just seen, warmer conditions make CO_2 less soluble in water, causing the oceans to give up CO_2 to the atmosphere as the temperature increases, a process that takes 600 to 800 years.

According to IPCC climate modelers, the melting of ice sheets and glaciers caused by the slight initial warming could not have continued, unless this temperature rise was amplified by positive feedbacks – including CO_2 feedback,

TABLE 5.3: CO_2 SENSITIVITY

The IPCC position

- The major CO_2 feedbacks, which control the climate's response to added CO_2 in the atmosphere, are *positive* feedbacks that amplify the effect of CO_2 on its own.

- The climate sensitivity, which measures any global warming from CO_2, has remained unchanged from the time of the ice ages until now.

- Global warming at the end of ice ages was aided by positive CO_2 feedback, though the feedback lagged the temperature by 600 to 800 years.

- Today's global warming is also assisted by positive CO_2 feedback, but the feedback switches on almost immediately.

WHAT'S WRONG WITH THIS PICTURE?

1. Recent satellite observations show strongly *negative* feedback from clouds. If this negative cloud feedback dominates the positive feedbacks from water vapor, and from snow and ice, the net CO_2 feedback is negative – diminishing the warming from CO_2 to an insignificant level.

2. Climate conditions during the ice ages were radically different from today, so it's unlikely that the climate sensitivity is still the same, as assumed in most IPCC computer models.

3. CO_2 lagged temperature during post ice-age warming because of its delayed release from the oceans. But because of the lag, atmospheric CO_2 kept rising for 600 to 800 years after the temperature leveled out. In the modern era since 1850, the CO_2 level and temperature have increased together. This two-faced CO_2 behavior makes no sense.

triggered by the surge in atmospheric CO_2 as it came out of the oceans.[159] Aided by the feedbacks, says the IPCC, a period of global warming ensued, with the temperature climbing until it reached a new, higher equilibrium level that signaled the end of the ice age. A similar chain of events, based on CO_2 and other feedbacks, enhanced global cooling as the temperature dropped at the beginning of the ice age.[160]

The problem for believers in the CO_2 global warming hypothesis is that when the ice age was over, the temperature suddenly stopped increasing and leveled out, but CO_2 continued to rise for another 600 or 800 years before plateauing and then decreasing slightly. That's well-documented in the historical record (Figure 5.1).[161]

But how can rising CO_2 in the atmosphere be the cause of escalating temperatures today, yet not 11,000 years ago – and at previous ice-age terminations – when the mercury stood still as the CO_2 level kept ascending? If CO_2 from the oceans didn't make the temperature go up during the 800 years or so immediately after the Earth's recovery from an ice age, why should CO_2 from human emissions make it go up now?[162] Perhaps some of the present rise in the CO_2 level is the ocean-delayed response to distant medieval warming.

It makes no sense that CO_2 feedback should be dragging the temperature upward in our present climate, and did the same thing as the planet pulled out of past ice ages, but then suddenly turned off for the final 600 to 800 years.

The IPCC, in its 2007 report, implies that this two-sided temperature response to CO_2 occurs because temperature and CO_2 are going up much faster during current global warming than they did at the end of the last ice age.[163] The atmospheric CO_2 level is indeed growing a lot more rapidly today, but the claim that the temperature is increasing 10 times faster than in the past is an exaggeration, as you might expect from the IPCC. The warming we've seen since 1850 has in fact been only two to three times faster than post ice-age global warming.[164] That's not such a big difference.

Other questions also arise. For example, the IPCC asserts that net CO_2 feedback has always been positive, from the ice ages to the present. I've already discussed the possibility, based on recent satellite observations of clouds, that the net CO_2 feedback in our current climate is negative. How can we be so sure it wasn't negative instead of positive going into and out of ice ages?

And how can we be certain that CO$_2$ feedback was operating at all in glacial times? Other feedbacks – such as from snow and ice – are believed to have played a role in ice-age temperature swings. Perhaps these other feedbacks could have done the job alone, without any help from CO$_2$, the atmospheric CO$_2$ level moving up or down simply in response to the changing solubility of CO$_2$ in the oceans caused by changing temperatures.

Assuming that there actually was CO$_2$ feedback to amplify temperature rises and falls, why was the temperature descent at the onset of an ice age so much slower than its rapid climb at the end?

If you're beginning to think there's more uncertainty than certainty in global warming science, you're right. The CO$_2$ hypothesis and the whole theory of human-induced global warming, for a theory is what it is, are a flimsy house of cards,

AN UNLIKELY VILLAIN

The CO$_2$ global warming hypothesis maintains that global warming is caused by human CO$_2$ emissions, because temperatures and the amount of CO$_2$ in the atmosphere happen to be going up at the same time. To review why CO$_2$ has been made the scapegoat for climate change, let's revisit the three red flags I raised for the CO$_2$ hypothesis in Chapter 2 (see Table 2.1):

- *The CO$_2$ steady level problem*: If the CO$_2$ global warming hypothesis is valid, temperature and the CO$_2$ level should track one another over all periods of time. Changes in temperature should reflect changes in CO$_2$. They appear to, but only for the last 160 years and during the ice ages, not for the thousands of years in between that included the Medieval Warm Period and the Little Ice Age. The IPCC's attempt to rewrite history in order to rectify this discrepancy led to the now discredited hockey stick, as described in Chapter 3. And even during the ice ages, the temperature led, rather than followed, CO$_2$.

- *The CO$_2$ amplification problem*: On its own, atmospheric CO$_2$ can't account for global warming without positive feedback, to amplify the effect of CO$_2$ and to boost the climate sensitivity. If the net feedback in the Earth's climate system is negative instead of positive, the climate sensitivity is low, and any future temperature increase from added CO_2 will be insignificant. Recent satellite data indicate that negative cloud feedbacks may be strong enough to result in negative overall feedback.

- **The CO_2 lag problem**: When past ice ages terminated, the temperature remained steady while the CO_2 level in the atmosphere kept rising for 600 to 800 years, as CO_2 continued to escape from the oceans. This is incompatible with the CO_2 global warming hypothesis, which says that today's rising levels of CO_2 from man-made sources are causing temperatures to increase. Yet the IPCC uses ice-age climate data to validate its computer models, on which its future climate projections are based.

Chapter 6: Doing What Comes Naturally

As we've seen in the last few chapters, man-made CO_2 is highly unlikely to be the main cause of climate change. The whole IPCC case for CO_2 warming is shot through with problems, from shoddy handling of the raw data to corruption to overconfidence in computer models to roadblocks for the CO_2 hypothesis. The science behind the IPCC's reports is so badly flawed that it's hard for me to believe that so many accept them as an act of faith.

But if isn't CO_2, what is it that's heating up the planet?

An honest answer to the question would be that we just don't know right now. It's not that we're short of ideas, but we simply don't have enough data or evidence at the moment to draw any firm conclusions – except that the chances of CO_2 being the number one culprit are very, very slim.

If humans are not to blame, the obvious place to look for an alternative explanation is nature. It's well-known that natural cycles have altered the Earth's climate many times in the past, notably the orbital changes that sent the globe into the deep freezes of the ice ages.

Nobody is suggesting that our current warming spell has anything to do with orbital cycles, but there are several other candidates in the natural world. One, or a combination, of these natural cycles could well be responsible for global warming. In this chapter, we'll examine two of these possible alternatives that stand out from the rest: our sun, and a natural short-term climate cycle called the Pacific Decadal Oscillation.

NATURAL WARMING: THE SUN

It has been known that the sun is linked to the Earth's climate ever since the invention four hundred years ago of the telescope, which made it possible to see sunspots (and other astronomical quirks) more clearly. Sunspots are small dark blotches on the sun's surface caused by magnetic storms in the sun. During the so-called Maunder Minimum, a 70-year period in the 17th and 18th centuries that formed part of the Little Ice Age, there were hardly any sunspots at all.

The number of sunspots that show up over a year is one of several solar cycles. The annual sunspot number goes up and down over an interval of about 11 years, and the cycle duration fluctuates as well, from as short as 9 years to as long as 14 years.

Along with the number of sunspots, the sun's heat and light output[165] jiggles during the solar cycle. Although this variation is too small, at less than one part in a thousand, to have any appreciable effect on our climate, the average value of the sun's output changes slowly with time – and these changes can cause warming and cooling of the Earth's surface.

There are cycles longer than 11 years too. Not only do solar activity and the sunspot number repeat every 11 years or so, but their maximum (and minimum) values also make extra big jumps every 87 years and again at 210-year intervals. Various regional climate changes have been tied to the 87-year cycle.[166] And the 87-year and 210-year cycles may be associated with a newly discovered 1,500-year cycle.

As well as direct connections between the sun and the Earth's climate, there may also be indirect solar effects involving cosmic rays from outer space, the sun's own ultraviolet radiation, or even other natural climate cycles.

So the sun can affect our climate and contribute to global warming in several different ways. But it should come as no surprise that the IPCC's latest report attempts to minimize this solar influence, in order to maximize the role it claims is played by CO_2.

More IPCC Shenanigans: Solar Warming Minimized

The IPCC has done it again.

As we saw in an earlier chapter, the IPCC has been highly selective in choosing data to bolster its case for man-made global warming. That's what the panel did with the CO_2 record, cherry-picking the data for 19th century CO_2 levels (Figure 3.5) to make the present-day CO_2 increase look higher than it really is.

The same wiliness can be found in the IPCC's treatment of historical data for solar activity, except that this time the goal was to make the contribution of solar variability to global warming appear lower than it actually is. The sun's variations probably can't explain all the global warming to date, but the IPCC's Fourth Assessment Report in 2007 minimizes the solar contribution almost to the point of insignificance by cherry-picking the historical record.

Proxy records of solar intensity include both sunspot numbers and so-called cosmogenic isotopes, which are trace amounts of radioactivity left in proxies such as ice cores and tree rings by past cosmic rays from space. As the sun's output waxes and wanes, the radioactivity levels of these isotopes rise and fall, leaving an imprint of solar intensity at the time.[167]

The variation in solar activity includes short-term fluctuations during the sun's 11-year cycle, as well as longer-term increases or decreases in the average solar output that can have a significant effect on climate. How big an effect the sun is having on present global warming depends on how much the average amount of energy coming from the sun has risen above normal.

According to the IPCC,[168] the sun's output right now has increased only slightly over the 300 years since the Maunder Minimum, the period from 1645 to 1715 when there were almost no sunspots and solar activity was low – the period that coincided with the coldest episode of the Little Ice Age. If the IPCC is correct, it means that any solar contribution to global warming today is minimal.

However, the IPCC's conclusion about the sun's output is based on a very selective choice of historical data, namely the plucking out of one of the lowest available estimates of the boost in solar output since the Maunder Minimum – as well as possible tampering with recent satellite data, which I'll discuss shortly.

The low estimate of the increase in solar activity during the last 300 years is derived from recorded sunspot numbers and a computer model.[169] Apart from the fact that computer simulations of the sun are probably no more reliable than computer climate models, the reconstruction of solar output in this particular model goes back only to 1713, which barely includes the Maunder Minimum itself that ended in 1715.[170]

The IPCC's 2007 report actually lists both high and low estimates of the solar activity gain since the Maunder Minimum, the high estimate being three times larger than the low one based on sunspot proxies, but calculated instead from ice-core radioactivity levels.[171] But the high estimate is cherry-picked too, as the IPCC neglects other even higher estimates from the same source.[172]

Needless to say, the IPCC then goes on to ignore all but the lowest estimate to calculate the solar contribution to global warming.[173] Obviously, a bigger increase in the sun's average output since the end of the 1600s would account for more of our current global warming than the IPCC wants to admit.

In its 2001 Third Assessment Report, the incorrigible IPCC resorted to a different subterfuge to minimize the sun's effect on global warming.

Most of the estimates in that report weren't based on proxy data from sunspots or ice cores, but on comparisons between the sun and similar stars, a method that is now largely discredited. But these comparisons yielded calculations of the solar activity gain since the Maunder Minimum even larger than the high estimate that the IPCC rejected for its 2007 report. To lower the big difference in solar output between past and present, the IPCC deviously measured the present-day increase only from the year 1750[174] – well after the Maunder Minimum was over, and the sun's intensity and the number of sunspots had returned to more normal levels. This ruse cut the solar activity increase since 1700 in half.

Once again, and on two separate occasions, the IPCC has cast aside the rules of science to prop up its unconvincing case for man-made CO_2 as the source of global warming. As I discussed earlier, one of the cornerstones of the scientific method is that you don't ignore or discard data just because it doesn't fit your theory. Properly conducted science examines all the data, without bias.

So determined has the IPCC been to trivialize the role played by the sun in global warming that it doesn't even bother to enumerate the solar contribution in its 2007 report. It appears that the IPCC estimates the solar portion of total global warming at only a few percent.[175]

Unfortunately for science, that's not the end of the story.

How much the sun's energy output has gone up over the 300 years since the Maunder Minimum depends not only on the historical record, but also on accurate knowledge of today's solar activity, which comes from satellite measurements. According to the IPCC,[176] the average solar activity level has remained the same during the last two 11-year cycles, based on one of the available sets of satellite data.[177]

But another set of satellite data shows a distinct increase in average solar output over this period,[178] which means that the solar activity gain from the time of the sunspot-free Maunder Minimum is bigger yet than estimated – and that the sun's contribution to global warming is much higher than the IPCC is saying.

What is disturbing about the IPCC's position is that Richard Willson, who is one of the managers of the second satellite data set, has recently accused his counterpart for the other set of satellite measurements of "unauthorized and incorrect adjustments" to the data, which shows nearly constant solar output since 1978.[179, 180] This accusation of data tampering, backed up by other solar scientists, is another example of the lengths to which the IPCC will go to counter any evidence that contravenes its assertions about CO_2 and higher temperatures.

There is plenty of evidence that solar activity has increased considerably over the last 300 years, and that the sun's role in global warming has been underrated in IPCC computer climate models.

Atmospheric researcher Joanna Haigh cites several studies, all of which suggest that climate models "may underestimate solar influence by up to a factor of three".[181] A study by mathematical physicists Nicola Scafetta and Bruce West, based on a completely different type of model, concludes that the sun may have contributed more than 50% of the observed global warming since 1900.[182]

The IPCC of course ignores all these studies. As an outside observer, I've found it hard to avoid concluding that the IPCC's climate modelers have intimidated most of the solar science community into accepting the IPCC view that the sun has little to do with present global warming, despite all the data to the contrary.

The 1,500-year Climate Cycle

A recent discovery that caused quite a stir in climate science circles was a natural cycle approximately 1,500 years long that manifested itself during the Earth's ice ages. This 1,500-year climate cycle is now believed to continue through the warmer interglacial periods as well, including the one we're in today.

The original discovery was made in the 1980s by paleoclimatologists Willi Dansgaard and Hans Oeschger, who examined ice cores obtained by deep drilling the Greenland ice sheet.[183] What they found was a series of rapid climate fluctuations, when the icebound Earth suddenly warmed slightly over a period as short as decades, only to gradually cool down again to frigid glacial temperatures.

These abrupt warming spells, of which 25 have been identified during the most recent ice age, are known as Dansgaard-Oeschger events. If you look closely at the ice-age temperature data in Figure 5.1, you may be able to identify some of them.

After the Greenland ice-age cycle was also discovered in Antarctic ice cores, further evidence from seabed sediment cores revealed that the 1,500-year cycle had persisted from the end of the last ice age right up until the present.[184] Buried in the deep-sea cores were the signs of a millennial-scale climatic cycle that keeps going like the Energizer Bunny, periodically nudging the temperature up and down during both glacial and interglacial epochs. The climate changes are equally abrupt whether the Earth is in an ice age or not.

But the cycle no longer repeats itself with the same clockwork regularity of the Dansgaard-Oeschger warmings during the ice ages. The more recent fluctuations have instead come and gone at irregular intervals, varying from about 1,000 to 2,000 years.[185]

The 1,500-year (plus or minus 500) climate cycle has been linked to the sun, as described at length in a recent book by Fred Singer and Dennis Avery.[186] The solar connection is backed up by a wide variety of proxy evidence that includes not only ice cores and seabed sediments, but also tree rings, pollen fossils, coral reefs, cave stalagmites, and advances and retreats in both glaciers and tree lines.

A problem for this explanation, however, is that there is no 1,500-year solar cycle. As I mentioned earlier in the chapter, the sun's natural variability includes 87-year and 210-year cycles, as well as the 11-year sunspot cycle, though nothing known to be of longer duration.

In order to explain Dansgaard-Oeschger events – the abrupt changes in temperature that punctuated the ice ages at 1,500-year intervals – a group of German climate researchers has proposed a model in which variations in solar activity do indeed pace an Earth cycle of close to 1,500 years, at least during ice ages.[187]

The basic idea is that a combination (or superposition, in mathematical terms) of the sun's 87-year and 210-year cycles could produce a terrestrial cycle 1,470 years in length. That's because seven of the longer 210-year solar cycles or seventeen of the shorter 87-year cycles fit exactly into 1,470 years, the two solar cycles giving each other a boost once every 1,470-year earthbound cycle.[188]

To test their idea, the researchers subjected a computer climate model to ice-age climate conditions. Their model, when "forced by periodic freshwater input into the North Atlantic Ocean in cycles of 87 and 210 years", successfully simulated rapid climate shifts similar to Dansgaard-Oeschger events, spaced 1,470 years apart.[189]

Although the German team did not attempt to simulate the less regular climate changes that have occurred over the 11,000 years since the last ice age finished, the scientists point out that there is ample evidence over this period of events associated with the 87-year and 210-year solar cycles that were used as climatic disturbances in their ice-age model.

In fact, close examination of the seabed sediment evidence for a post ice-age 1,500-year climate cycle shows that it really consists of cycles within a cycle, with several smaller "nodes" – corresponding to the solar 87-year and 210-year cycles – within each segment of the longer cycle.[190] A slightly different interpretation has recently been put forward, in which the 1,500-year cycle is not of solar origin but arises from circulation of deep ocean currents, while the sun drives two other related cycles, 1,000 years and 2,500 years long.[191]

With lots of reliable data showing that a climate cycle of 1,500 years (plus or minus 500 years) exists and that it's connected to the sun, the key question is whether the cycle can account for our present global warming.

If the 1,500-year cycle does explain higher temperatures today, then there should have been earlier periods of global warmth 1,000 to 2,000 years ago, and again 1,000 to 2,000 years before that. And there were: the Medieval Warm Period, which began around the year 800 (see Figure 3.2), and the so-called Roman Warming about 1,000 years before. Both of these are well-documented in the historical record.[186]

Nevertheless, the irregularity of the 1,500-year cycle's behavior makes it difficult to be certain that we're currently in another warm phase of the cycle. If the last warm phase kicked off in the year 800, the next one could have appeared as early as 1800 but, on the other hand, it may not show up for another few hundred years.

One check on this is the sun, since we already know that the 1,500-year climate cycle is tied to solar energy output. The research team that used deep-sea sediments to confirm the existence of the 1,500-year cycle also employed proxy data from ice cores and tree rings to connect the cycle to solar activity.[190] The same solar proxies, which involve long-lived radioactive isotopes, have been used by another research group to conclude that the sun's activity is higher today than it was during the Maunder Minimum, as discussed in the last section – although the authors say that the activity level has been as high as it is now for several brief periods over the last 1,000 years.[192]

This currently higher solar activity, which the IPCC goes out of its way to minimize, may be a sign that we've entered the next warm phase of the Earth's 1,500-year cycle. The IPCC, while not mentioning the 1,500-year climate cycle directly nor attempting to model it (nor the shorter 87-year and 210-year cycles), at least acknowledges the "millennial variability" of climate and its possible connection to the sun.[193]

Warmth from Outer Space

Variations in the sun's output are small, less than a tenth of one percent over the 11 years of the sunspot cycle, and perhaps only a few times as much during the 300 years since the Maunder Minimum ended in the early 18th century. But as we've seen, this may be enough to explain a substantial part of our current global warming, even though the IPCC wants us to believe otherwise.

How can such small changes in the sun's activity affect our climate at all?

One answer is in indirect solar effects – not the direct effect of the sun's heat, which is absorbed by gases in the atmosphere, by clouds, by the oceans, and by land surfaces;[194] but secondary, indirect effects associated with galactic cosmic rays, the sun's ultraviolet radiation, or maybe one of the Earth's atmospheric circulatory patterns. It's possible that a tiny increase in solar activity could be amplified by one of these indirect effects, through a positive feedback process.[195]

Cosmic rays are super-energetic, electrically charged particles that come mostly from exploded stars in our Milky Way galaxy and continually bombard the upper atmosphere. But the sun produces its own charged particles[196] that can deflect cosmic rays away from the Earth. As solar activity goes up, the number of cosmic rays hitting the atmosphere goes down; as solar activity declines, more cosmic rays get through to the atmosphere.

According to a recent proposal by physicist Henrik Svensmark, cosmic rays can seed the formation of clouds – especially low-level clouds that cover more than 25% of the Earth's surface and produce cooling.[197]

The idea is that a small increase in the sun's activity that decreases the number of cosmic rays can therefore reduce the cooling effect from low clouds, amplifying the direct warming effect from the sun itself. In Svensmark's words, the connection between clouds and cosmic rays provides "a mechanism for solar-driven climate change more powerful than changes in solar irradiance".[198]

Svensmark claims that this indirect cloud mechanism can explain not only modern global warming, but also the 1,500-year climate cycle discussed in the previous section, and even the onset and termination of ice ages in the very distant past. Millions of years ago, the variation of cloudiness induced by cosmic rays had less to do with solar activity, says Svensmark, than with changes in the number of cosmic rays reaching the sun in the first place, which fluctuated as the solar system moved across the spiral arms of our galaxy.[199] Other researchers have found that the number of cosmic rays in the sun's vicinity does indeed vary, over a cycle of about 140 million years.[200]

However, Svensmark's far-reaching assertions are largely speculative at present. The notion that cosmic rays are connected to low cloud cover is based on a small-scale laboratory experiment, which may not be a realistic simulation of actual conditions in the Earth's atmosphere.

A more definitive test of the cosmic ray contribution to global warming will come in 2010, when a full-scale experiment will begin at the European atom-smashing facility in Geneva.[201]

It's already well-established that the slight flickering in the sun's output regulates the number of cosmic rays that reach the atmosphere. If the Geneva experiment shows a link between cosmic rays and low clouds, it will mean that cosmic rays are a powerful amplifier of solar activity through their effect on cloud cover. Later in this chapter, we'll see how changes in cloud cover may cause global warming by a different mechanism involving the Pacific Decadal Oscillation.

The IPCC is hedging its bets on cosmic rays for now, saying that while "empirical associations have been reported between globally averaged low-level cloud cover and cosmic ray fluxes", its level of scientific understanding of cosmic ray influences is "very low".[202] This very low understanding has evidently prompted the IPCC to omit all indirect solar effects from its computer models – although its marginally higher "low" understanding of direct solar effects[203] hasn't prevented it from including these in the models.

Ozone Heating

Another indirect heating effect of the sun implicates the ozone layer.

You've probably heard about the holes in the ozone layer, which are a concern because they allow more of the sun's harmful ultraviolet (UV) radiation to penetrate the atmosphere and make it to the Earth's surface in those regions of the globe. The ozone layer normally protects us from most UV rays by absorbing them.

As the sun's output of heat and visible light fluctuates, so too does its production of invisible UV. A more active sun generates more UV radiation, which creates more ozone in the atmosphere.[204] The slightly thicker ozone layer provides a little more UV protection for humans by absorbing more sunlight.

Absorption of solar UV also warms the ozone layer, since all forms of radiation from the sun (including visible light, UV and infrared) carry heat energy. This in turn heats the atmosphere and amplifies the direct warming caused by solar activity,[205, 206] or so it was thought until recently.

However, this explanation for indirect heating has been questioned,[207] and other indirect solar mechanisms for the amplification of the sun's variations – in addition to cosmic rays and UV radiation – have been suggested.[208]

Other Planets

If the sun is playing a part in global warming on Earth, we would expect the other planets in our solar system to show signs of warming as well. But while there have been isolated claims that planetary global warming indeed exists, at this stage I believe all such claims should be taken with a pinch of salt.

One reason is the extreme scarcity of data and scientific studies. Even with the annual publication of hundreds of research papers on the climate of our own planet, there's plenty of disagreement about what is causing temperatures to rise – despite the constant efforts of the IPCC and global warming alarmists to convince us that CO_2 is the offender. So we shouldn't pay too much attention to what are just a handful of articles on warming trends elsewhere in the solar system.

An example of how easy it is to be misled by limited data can be found in recent studies of Mars, where dust storms are prevalent.

From darkening of the planet's surface, a team of space scientists calculated that in the 22 years between two NASA missions to Mars, it had warmed by 0.65° Celsius (1.2° Fahrenheit). This is about as much global warming as we've experienced here on Earth in 160 years. Attributing the warming to changes in sunlight-reflecting surface dust (though not to solar activity), the research team linked the sudden temperature rise to melting of the ice cap near the Martian South Pole over the past few years.[209]

But a different group of investigators pointed out that the first mission to Mars had come right after a global dust storm there that had made the planet appear much brighter than it looked 22 years later.[210] In fact, the surface brightness doesn't change a whole lot over time, so it doesn't tell us much about Martian warming.

The melting of the southern ice cap certainly could be another sign of the enhanced solar activity, discussed earlier in the chapter, which is currently influencing the Earth's climate. But, in the absence of any other data, it doesn't make a strong case for global warming on Mars.

Another reason to be skeptical about evidence for global warming on other planets is that most planets have very long "years". It takes 165 Earth years for Neptune to orbit the sun once, for example, which means we need to wait a long time to ensure that any warming trend we notice isn't just from the changing Neptunian seasons. Possible global warming reported on Neptune,[211, 212] on Neptune's moon Triton,[213] and on Pluto[214] have all been ascribed, at least partly, to planetary seasonal effects.

One planet where CO_2 has unquestionably caused global warming is Venus, which has a dense atmosphere consisting of 97% CO_2, far above the Earth's current CO_2 level of about 390 parts per million (0.039%). The runaway greenhouse effect on Venus produces furnace-like conditions, with a surface temperature around 470° Celsius (880° Fahrenheit). But that's how it's been there for billions of years and the planet isn't currently getting any warmer from the sun.

NATURAL WARMING: THE PACIFIC
DECADAL OSCILLATION (PDO)

Natural climate variability includes several known patterns of cyclic behavior. These come about through changes in wind and ocean circulation over months or years, but reveal themselves as fluctuations in temperature, rainfall and other features of our daily and weekly weather.

The familiar El Niño and La Niña cycles arise from seesaw changes in tropical ocean surface temperatures that cause drastic shifts in climate around the Pacific Ocean,[215] for periods of a year or more at a time. The 1997-1998 El Niño, one of the strongest on record, raised surface temperatures around the world by 0.17° Celsius (0.3° Fahrenheit)[216] – which is an appreciable fraction of the total global warming since 1850 – for a full year. More serious effects of El Niño and La Niña can range from catastrophic flooding in the U.S. and Peru to severe droughts in Australia. The cycles recur on a regular basis, although the intervals between them can vary quite a bit.

Other natural cycles that repeat at intervals from years to decades include the Arctic Oscillation, the Pacific Decadal Oscillation, and the Atlantic Multidecadal Oscillation. Although IPCC computer climate models endeavor to simulate the various cycles, the models have been unsuccessful at predicting the timing and climatic effects of several of them.[217]

Recently, Roy Spencer and his research team at the University of Alabama in Huntsville have proposed that *cloud* changes associated with the Pacific Decadal Oscillation (PDO) may be able to account for much of measured global warming,[218] which the IPCC and global warming alarmists insist on attributing to CO_2. The University of Alabama group is the same one that found evidence in satellite observations for negative CO_2 feedback from clouds, a subject discussed in Chapter 5. The new global warming theory also relies on satellite data.

The PDO has characteristics in common with the El Niño cycle, but the cycle time is much longer – about 30 years for the PDO, compared with the one to two year duration of El Niño. Although the fluctuations of the PDO can be traced back at least several centuries, its distinctive pattern wasn't recognized until the 1990s, when it was named by a U.S. fisheries scientist trying to explain the connection between Alaskan salmon harvests and the Pacific climate.

TABLE 6.1: NATURAL WARMING – ALTERNATIVES TO CO_2

Alternative

1. *Solar variability*
   - A 300-year increase in the sun's heat and light output since the Maunder sunspot minimum during the Little Ice Age.

   - The 1,500-year climate cycle, which has continued through both ice ages and interglacial periods, and may be tied to solar 87-year and 210-year cycles.

   - Indirect heating from the sun's blocking of cosmic rays. This blocking may reduce the cooling from low clouds when the sun is more active.

   - Indirect heating by the sun's UV radiation, which warms the ozone layer in the atmosphere.

2. Cloud fluctuations associated with the *Pacific Decadal Oscillation (PDO),* a natural climate cycle that switches between warm and cool every 30 years.

WHAT'S WRONG WITH THIS PICTURE?

- The IPCC claims erroneously that nearly all global warming is caused by man-made CO_2, and that natural climate variability plays almost no role.

- The IPCC wrongly assumes that the only natural contribution to the current warming trend is direct solar effects, and deceptively minimizes even this contribution.

- Potential sources of global warming, such as indirect solar effects and natural climate cycles like the PDO, are omitted from the IPCC's computer models.

- Most of the IPCC's efforts go into shoring up its case for the faulty CO_2 global warming hypothesis. Very little attention is paid to investigating alternative explanations for higher temperatures.

The PDO cycle alternates between warm and cool phases. During the 30-year warm phase, ocean surface temperatures in the northeastern Pacific are higher than normal, and the southeastern U.S. is cooler and wetter than usual; the 30-year cool phase is dominated by correspondingly lower ocean temperatures and warmer, drier conditions inland.

Long before Spencer's team made their satellite observations, meteorologist Joe D'Aleo[219] and others pointed out that the warm and cool phases of the PDO had coincided with the major periods of warming and cooling, respectively, in the 20th century. Even though the overall trend in temperatures since 1850 has been upward, there have been at least two intervals when the mercury took a dive for a decade or more (Figure 2.1). Each time that the PDO mode shifted from warm to cool, or vice versa, global temperatures switched accordingly (see Figure B.1).

To some skeptics, this coincidence explains global warming. But while the PDO may be an explanation for global warming, it is not simply because the temperature follows the warm and cool phases of the oscillatory cycle.

What Spencer postulates is that the natural variability of the PDO causes changes not only in temperature and precipitation, but also in cloud cover. As you may recall from earlier chapters, low-level clouds cool the Earth's surface, while high-level clouds have a heating effect. Warming can originate from either a reduction in low clouds, or an increase in high clouds, or both. In Spencer's words,

> Such a cloud change would cause the climate system to go through natural fluctuations in average temperature for extended periods of time. The IPCC simply assumes that this kind of natural cloud variability does not exist, and that the Earth stays in a perpetual state of radiative balance that has only been recently disrupted by mankind's greenhouse gas emissions.[218]

Disputing this IPCC claim, Spencer says that his group has already shown theoretically that random variations in cloudiness on a *daily* basis can cause ocean temperatures to vary over *decades*, which is one of the signatures of the PDO.[220]

The new University of Alabama proposal extends this idea to variations in cloud cover that occur not daily, but over timespans of up to 100 years – which might then cause long-term temperature changes such as global warming. Although the PDO is localized in the Pacific Ocean, the mixing of ocean waters around the world over a period of time can result in global cloud fluctuations.

Using a simple climate model with very few assumptions, and with fluctuations in cloud cover directly related to the observed PDO undulations, Spencer simulated global temperatures for the 20th century.[221] Astonishingly, this simple PDO model, on its own, can explain up to 75% of the global warming observed for the whole period from 1900 to 2007.

The other 25% could come from another natural source such as the sun, from CO_2,[222] or possibly from some combination of both. Some researchers have suggested that solar variability could affect the Earth's climate indirectly by perturbing natural cycles such as the Arctic Oscillation,[223] the North Atlantic Oscillation, or even the PDO.

A summary of the satellite evidence for PDO cloud fluctuations is given in Appendix B. Not surprisingly, analysis of the satellite data shows strongly negative cloud feedback, just as found in the earlier studies of other cloud data by Spencer's team.[224]

Negative feedback diminishes the warming caused by atmospheric CO_2 acting alone, to a low or even an insignificant level. As Spencer remarks, "The evidence continues to mount that the IPCC models are too sensitive, and therefore produce too much global warming. If climate sensitivity is indeed considerably less than the IPCC claims it to be, then increasing CO_2 alone can not explain recent global warming."[218]

In 2008, the 30-year PDO warm phase that began in 1977 came to an end. This ushered in a new cool phase, just as the shift that occurred in the early 1940s resulted in global cooling for the 30 years that followed. It's important to note that these cooling spells are also predicted by the PDO cloud fluctuation model, the basic idea being that the PDO causes long-term variability in cloud cover that leads to global warming and cooling.

Of course, a single study doesn't make a compelling case for PDO cloud fluctuations causing 75% or more of the global warming that we've seen so far. But, since the IPCC and global warming alarmists are most likely wrong about

CO_2, the PDO theory is certainly a good candidate for an alternative, natural explanation. Further investigation may well validate this alternative theory.

How very different these natural explanations are from the CO_2 story pushed by the IPCC and global warming alarmists! According to the CO_2 hypothesis that we've examined at length in previous chapters, nearly all the warming measured since 1850 has been caused by human emissions of CO_2.

Natural cycles, such as the PDO cloud changes discussed here, have been summarily dismissed by the IPCC – which makes the sweeping assumption that the only source of natural climate variability that contributes to global warming is solar activity.[225] And it resorts to deception to minimize even this contribution.

Admittedly, our current understanding of natural variability is poor. But that's no reason to exclude natural causes, other than the sun, from computer climate models.

The IPCC says its computer models are unable to match the measured temperature record for the 20th century if the models include only natural sources of warming.[226] That's because the only natural sources simulated by the models are direct solar effects, not to mention the intrinsic limitations of the models themselves. All other natural possibilities, including indirect solar effects such as cosmic rays, plus climate cycles like the PDO, are omitted.

The IPCC makes much of its supposed ability to detect an "anthropogenic signal" in climate data, especially the temperature record, as substantiation of man-made global warming.[227]

All this means in reality is that computer climate models, with all their assumptions about CO_2 and the climate system, can be adjusted to simulate the climatic record – and that's not at all surprising, with so many variable parameters to play with. The apparently good fit to climate data doesn't validate the models nor authenticate the connection between CO_2 and global warming. An equally good match to the data could be achieved with a different set of assumptions, based on a larger contribution from the sun and on natural cycles that the IPCC has little idea how to model.

An adequate understanding of natural climate variability is sadly lacking. For the last 20 years, the climate science community in general, and the IPCC in particular, has been obsessed with CO_2. If just a small fraction of the countless hours and weeks of effort that have been wasted trying to justify the CO_2 theory had instead been devoted to possible natural causes of rising temperatures, we might by now have had enough data to properly evaluate the alternative explanations.

Chapter 7: Global Cooling

One thing that both skeptics and alarmists agree on is that the CO_2 level is getting higher and higher, year after year. To alarmists, this relentless CO_2 increase is cause for panic, since temperatures have been rising too – which surely means that it will just become warmer and warmer. If we don't cut back on CO_2 immediately, goes the alarmist argument, then the temperature will reach a tipping point and there'll be no going back.

A big problem with this line of reasoning, apart from the faulty hypothesis linking CO_2 to global warming, is that the temperature is refusing to cooperate. From as far back as 2002, global temperatures have been falling.

That's right, it's been cooling.

REVERSAL OF FORTUNE

Despite alarmist protestations about the veracity of reports on the cooling trend ever since it first became evident, the temperature downturn gained credibility early in 2009 with the release of a study by two mathematical scientists at the University of Wisconsin-Milwaukee.[228] The study found that temperatures around the world, which had been rising steadily for the previous 30 years (see Figure 7.2), have leveled off and have actually decreased slightly since late 2001.

And as if that wasn't enough to dismay global warming alarmists, the study predicted that the cooling will continue for up to 30 years from now. Stating that the cool spell can't be attributed to any particular cause, although it must be natural, co-author Kyle Swanson remarked, "This is nothing like anything we've seen since 1950".[229]

But this hasn't stopped the amount of CO_2 in the atmosphere from escalating. Figure 7.1 displays temperatures since 2002, measured by satellite,[230] together with the CO_2 level. Temperatures from Earth-based thermometers reveal identical behavior to the satellite data over the same period.[231]

The plunge in temperature, shown by the trend lines in Figure 7.1, is unmistakable. Up to early 2009 (which is not included in the graph), the average temperature has tumbled by almost 0.15° Celsius (0.3° Fahrenheit) over slightly more than seven years.[232] That cancels out almost a third of the gain since the present warming spurt began in the 1970s.

Figure 7.1: Recent Temperatures and the CO_2 Level

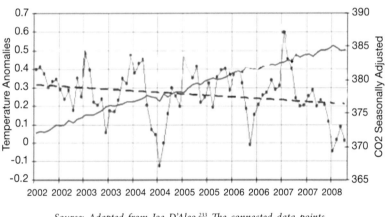

Source: Adapted from Joe D'Aleo.[233] The connected data points show the satellite temperature record (at monthly intervals) since 2002, while the solid line is the CO_2 level measured at Mauna Loa.

You'll be wondering, no doubt, what the IPCC and climate alarmists have to say about all this. Fond of accusing skeptics of being climate change deniers, alarmists at first were in denial themselves about falling temperatures. But once the reality of the current cooling trend became apparent, the alarmist movement adopted a new ploy.

Which is that cooling is the new warming!

I'm not kidding. There are lots of folk out there, scientists and nonscientists alike, who believe global cooling simply proves that global warming exists. Never mind how illogical this is. As recently as December 2008, an Associated Press correspondent wrote, in a story carried by ABC News and other major news media:

> Ironically, 2008 is on pace to be a slightly cooler year in a steadily rising temperature trend line. Experts say it's thanks to a La Niña weather variation. While skeptics are already using it as evidence of some kind of cooling trend, it actually illustrates how fast the world is warming.[234]

The mental contortions of this kind of thinking aside, we should recognize that an extended period of cooling – even if it lasts for another 30 years – isn't necessarily the end of global warming. Shown below (Figure 7.2), and also in Chapter 2, is the temperature record from 1850, which is roughly when modern global warming is considered to have begun. As you can see, there have been several cooling stretches over this time, during which the mercury went up about 0.8° Celsius (1.4° Fahrenheit) overall.

The chart in Figure 7.2 can be divided into five time periods of close to 30 years each, in which the temperature alternately rose and fell. From 1850 to 1880 the temperature increased; from 1880 to 1910 it went down again, only to go up once more until 1940; and so on.[235] In fact, this alternating pattern in the temperature record goes back all the way to 1470, with a cycle time of around 27 years.[236]

That 30 years is the cycle time for the Pacific Decadal Oscillation discussed in the previous chapter, and also approximately for the Atlantic Multidecadal Oscillation,[237] may be significant, though our knowledge of both these natural cycles is much too limited at present to draw any conclusions from this observation.

What will happen when the recent dip in temperatures comes to an end is anyone's guess. Based on the 30-year historical pattern just discussed, we'd expect another bout of warming. Indeed, the University of Wisconsin study concludes that the present cooling is "superimposed upon an overall warming presumably due to increasing greenhouse gases".[228]

But this assumes that the underlying climate driver is CO_2, an assumption that is very likely wrong as we've seen. And even though the study authors profess allegiance to the CO_2 hypothesis, that may be just for the purpose of maintaining their standing in the climate science community.

As discussed in Chapter 5, the sensitivity of the Earth's climate system to added CO_2 is unlikely to be anywhere near as high as the IPCC claims. The very fact that we're currently in a cooling period while CO_2 is still rising attests to low CO_2 sensitivity.

Figure 7.2: The Temperature Record from 1850

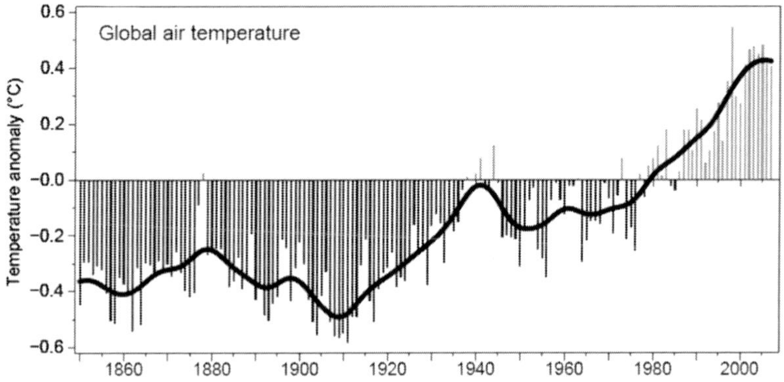

Source: Climatic Research Unit (CRU). [238]

So it's not at all certain that temperatures will resume their previous ascent when the natural source responsible for the present decline switches off in 20 or 30 years. It's entirely possible that some other source of natural variability will kick in and keep the temperature trending down.

Some solar scientists have predicted that the sun's energy output, currently believed to be at a 300-year high as discussed earlier, will diminish over the next two decades, and that by 2030 the reduction in solar activity alone will result in substantially cooler temperatures – perhaps even comparable to the chill of the Little Ice Age in the 17th century.[239]

We just don't have enough knowledge of the climate to make a reliable prediction one way or the other, to say whether it will warm, or cool further, when the present cooling is over. The IPCC and global warming alarmists have brainwashed us into believing that the steady increase in temperatures over the latter part of the 20th century will continue unabated. A quick glance at the long-term historical record (Figure 3.2), however, tells us that global temperatures are always going up and down over short periods of time.

CRYING WOLF

Global cooling, even for a limited period, isn't in the IPCC playbook. If not hell on earth, something close to it is projected by the IPCC's most recent report, with ever climbing global temperatures from now until the end of the 21st century and beyond. Along with the nonstop warming, says the report, will come intensified heat waves, surging sea levels, and prolonged droughts.[240]

All these projections are based on the IPCC's computer climate models – which, as we know, rely heavily on CO_2 (and other greenhouse gases), with a very small contribution from the sun but not from any other natural sources. The predictions encompass a range of futuristic scenarios that take into account varying estimates of CO_2 emissions, global population and economic growth.[241]

While the various scenarios result in different predictions of the assumed temperature leap by century's end, the warming calculated for the first few decades until 2030 is almost the same in all pictures. From 2002 to early 2009, the global warming projected by the IPCC was about 0.14° Celsius (0.25° Fahrenheit).[242]

How wrong can you be?

Over these seven years, average surface temperatures have *dropped* by as much as the IPCC's predicted gain for the same interval. This means the IPCC overestimated the temperature for 2009 by close to 0.3° Celsius (0.5° Fahrenheit).[243] If the cooling trend continues as it looks like it will, by 2011 or 2012 the total temperature increase since 1980[244] – which is half that for the whole period since 1850 – will have been wiped out.

Consumed by recent high temperature readings that were the highest in hundreds of years, the IPCC failed to even notice the cooling downturn. Its 2007 report includes recorded temperature data to as recently as 2005, three years into the slump, and yet its projection graphs show the temperature inching up every year from 2000.[242] Talk about not having your act together.

The IPCC's temperature estimates for the first decade of the century are so far from reality that it's impossible to give any credence at all to its inflated projections for 2100, projections in which the IPCC claims to have a confidence level of 66%.[245]

And it's not just temperature predictions on which the IPCC and alarmists in general are crying wolf. The list of highly questionable IPCC projections for our future climate includes heavier rainfall, more intense hurricanes, and widespread thawing of permafrost.[246] There's little likelihood that any of these will occur if the climate is dominated not by CO_2 emissions, but by natural variability that is currently cooling the globe down.

Global warming alarmists made hay of Hurricane Katrina, the deadly hurricane that struck the U.S. Gulf Coast in 2005, causing widespread flooding and devastation. It was all too easy to attribute such a costly disaster to global warming. Indeed, the IPCC maintains that "intense tropical cyclone activity has increased since about 1970", and that tropical cyclones (which include hurricanes and typhoons) are now much stronger and longer in duration than before, due to global warming.[247] The IPCC also predicted that future tropical cyclones will become more intense yet, though the total number is not expected to increase.[248]

However, hurricanes and tropical storms are currently becoming less intense, if anything. A commonly used measure of tropical cyclone strength and duration is now at its lowest point since the mid-1970s (Figure 7.3). This almost certainly reflects recent global cooling, just as the previous upward trend in cyclone activity noted by the IPCC, and visible in Figure 7.3, was indicative of the warming spurt that ended in 2001.

Figure 7.3: Tropical Cyclone Activity since 1977

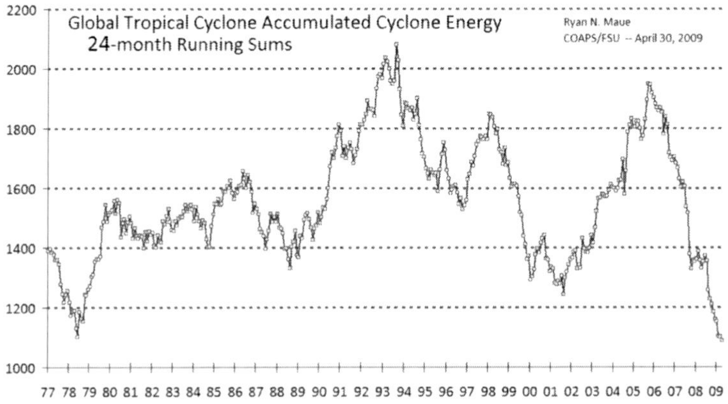

Source: Maue.[249]

A possibly more pressing issue, often in the news recently, is the extent of Arctic sea ice. For years, those who adhere to the theory that CO_2 causes global warming have been lamenting the retreat of glaciers, the disappearance of snow caps from

TABLE 7.1: GLOBAL COOLING

The IPCC's stand

- Global temperatures will continue to increase steadily through the end of the 21st century, unless we rein in our emissions of CO_2.

- Consequences of global warming such as more intense hurricanes, shrinking of the Arctic ice cap, and retreating glaciers, will get worse without action on CO_2.

- The oceans will continue to warm, causing sea levels worldwide to rise drastically.

WHAT'S WRONG WITH THIS PICTURE?

1. Man-made CO_2 is highly unlikely to be the principal cause of global warming, as discussed in previous chapters.

2. Since 2002, global temperatures have been falling. A new study predicts that the cooling trend may continue for another 30 years.

3. The IPCC's most recent report doesn't predict any cooling at all, and overestimated the projected temperature for 2009 by 0.3° Celsius (0.5° Fahrenheit) – which is almost half the total global warming since 1850.

4. Average hurricane strength and duration are currently at exceptionally low levels, reflecting the global cooling trend.

5. Although Arctic sea ice shrank from 1979 to 2005, it has begun to expand again, probably due to global cooling .

6. A 50-year warming of the world's oceans ended in 2003. Since then, the oceans may have cooled slightly, and the rate of increase in sea level has slowed.

lofty mountains, and the shrinkage – from the early 1970s to about 2006 – of the ice cap at the North Pole. If only we'd stop putting more CO_2 into the atmosphere, they insist, all of this could be reversed.

The shrinking of the polar ice cap has in truth been quite dramatic. Satellite measurements show that Arctic ice cover in the summer months, when the ice is at its minimum extent, contracted at a rate of 7.4% per decade from 1979 to 2005, which is a total reduction of 20% over that period.[250]

You don't have to be a rocket scientist (or even a climate scientist) to figure out that, if the same trend continued, the ice cap wouldn't be here much longer. In 2008, an ice scientist declared that Arctic ice was "in a death spiral" and could be completely gone in summer by about 2030.[251] No more ice to fend off the global warming rays of the sun's summer heat,[252] no more habitat for polar bears.

But a funny thing happened on the way to disaster. Since at least 2007, and perhaps a year or two earlier, Arctic ice has been expanding again.[253] The summer ice cap in 2008 was 9% larger than the minimum reached in 2007, the minimum that had sent alarmists into such a frenzy. In other words, the gain in just one year was greater than the loss over a whole decade during the 1980s and 1990s.

Of course, the recovery of the ice cap that appears to be underway is probably linked to the cooling trend discussed in the previous section. Global warming alarmists want none of it, however. Recent press releases have reported only that Arctic ice cover was at its "second lowest" or "third lowest" extent ever in the last few years[254] – supposedly providing further evidence of the 30-year-plus decline, even as the ice has started growing once more.

Global cooling, it seems, is here to stay for a while. It will be interesting to see what distortions of the truth the IPCC comes up with in its next assessment report (due in 2014) in order to explain so many failed predictions about global warming.

OCEAN WARMING ON HOLD

When the global warming music stopped, it wasn't just Arctic sea ice that took notice. The oceans have stopped warming as well.

Oceans play an important role in regulating global temperatures because they can hold a lot more heat – about 1,000 times more – than the atmosphere. More than 80% of the heat that global warming has added to the Earth's climate system over the last 50 years is sitting in the oceans.[255] The oceans can also hold much more CO_2 than the atmosphere, as we saw earlier. Both heat and dissolved gases such as CO_2 are initially stored in the upper layers of the ocean, where they can be readily exchanged with the air above.

Global warming or cooling shows up in the oceans in two ways. First, sea surface temperatures rise or fall, just as surface temperatures do on land. Second, the total heat stashed away in the ocean depths changes with temperature, more heat being stored as the Earth warms and less as it cools.

The cooling trend that began in 2002 is visible in both land and sea surface temperatures, though these are usually combined to produce records like those in Figures 7.1 and 7.2. The same downturn in temperature, delayed by two years, was initially reported in ocean heat storage as well.[256]

However, it turned out that the report about diminished ocean heat content was not completely correct, because of bias in the underwater robots used to measure the heat.[257] The robotic gadgets are buoys that patrol the world's oceans, sinking more than a mile deep once every week or so and then bobbing up to the surface, taking the temperature of the water as they ascend. When the floats eventually reach the surface, the data is transmitted to a satellite.[258]

After correcting for the bias of the faulty instruments, the authors of the original study concluded that the heat stored in the oceans has leveled off, rather than gone down, saying that no significant warming or cooling of the upper ocean has occurred since 2003.[259] Nevertheless, other studies based on the same corrected data find that there has indeed been a slight falloff in ocean warmth over the last six years.[260, 261]

Either way, it's clear that the oceans have stopped warming, at least for now. This happened once before for a short period, in the early 1980s, although the overall trend in stored ocean heat over the 50 years to 2003 was unquestionably upwards.[262]

But the previous interruption, when the oceans cooled very quickly, lasted only five or six years before the warming resumed.[263] So we should know within a year or two whether the present hiatus in ocean warming is temporary like before, or whether the watery depths are going to follow in the footsteps of surface temperatures and cool off also. Every indication is that global warming – of land masses, the oceans, and polar ice – is over for now.

The IPCC said little in its 2007 report about ocean cooling, except to mention that after the oceans had warmed by 0.1° Celsius (0.2° Fahrenheit) from 1961 to 2003, there had been "some cooling".[264]

Sea Levels

Ocean warming or cooling has important ramifications for sea levels. Since water expands as it warms, higher ocean temperatures raise the average sea level; conversely, lower temperatures bring the level down. Sea levels also rise as glaciers and ice caps melt, but by far the biggest contribution currently comes from thermal expansion of the oceans.[265]

The 50-year warming of the Earth's oceans and the sudden five-year cooling in the early 1980s are both reflected in the globally averaged sea level, as documented by the IPCC.[266] Over this 50-year period, the seas rose about 100 mm (4 inches). And the 10 years from 1993 to 2003, when global temperatures were escalating rapidly (see Figure 7.2), saw a correspondingly higher rate of increase in the sea level.

But recent satellite data for the average sea level shows that this rate of increase slowed down after 2003[267] – suggesting that sea levels are behaving in the same way as ocean heat storage which, as we've just seen, has either flatlined or declined in the last six years.

The prospect of massive flooding of low-lying coastal areas and islands, from the IPCC's projected rise in sea levels arising from global warming, has captured the public imagination and the attention of the media like no other potential change in our climate. But as with future warming and the melting of Arctic sea ice, the IPCC and other alarmists may once more be crying wolf.

Chapter 8: Why It Matters

Almost everything in the book so far has been about the *science* of global warming. That's because the IPCC's reports on global warming – which have become the manifesto for climate change alarmists – are founded on flawed science and abuse of the well-established scientific method. I've shown just where the IPCC has gone astray scientifically in trying to tie global warming to man-made CO_2.

But, the science aside, why does it matter?

Whether or not global warming comes from CO_2 is important because the human race has embarked on what could become one of the costliest efforts ever undertaken, just in order to curtail CO_2 emissions. The effort could cost the U.S. alone up to $2 trillion,[268] which is half as much as the country spent on World War II.[269]

Trillions of dollars could be squandered, to fix a problem that doesn't exist. Global warming may be real, but there's next to no evidence that CO_2 has very much to do with it.

Nevertheless, global warming alarmists are pushing hard for measures to reduce our carbon footprint such as emissions caps, tradable carbon credits, and sequestering of CO_2 in underground reservoirs. Corporations have jumped on the carbon bandwagon too. And legislative bodies around the world are getting into the act, beginning with the UN Kyoto Protocol to limit production of CO_2 and other greenhouse gases.

Carbon regulation is already upon us. But as recent European experience has shown, no amount of regulation, whatever its intent, can guarantee lower CO_2

emissions. And even if CO_2 discharges were to drop, it probably wouldn't make any difference whatsoever to global warming.

THE CARBON CAP BOONDOGGLE

According to alarmists, human CO_2 emissions are a serious threat to our planet and need to be scaled back before the climate reaches a point of no return. Mistaken though this notion is, the alarmist community long ago realized that curbing CO_2 would require intervention of some kind. The biggest sources of CO_2 are power plants and smokestack industries, neither of which is inclined to take voluntary action on an invisible gas that has no known health effects at its normal atmospheric concentration.

Two main approaches can be taken to controlling CO_2 emissions into the atmosphere: regulatory limits that apply to all emitters across-the-board, regardless of the quantity of CO_2 released, and market-oriented methods such as emissions trading or a carbon tax.

Because straight regulation is not seen as a very cost-effective approach, given its need for an extensive bureaucracy and an inspection system, interest usually focuses on a market-based alternative – either an emissions tax, or a trading scheme often known as cap-and-trade. Various cap-and-trade systems already exist around the world, on a national or regional level, to limit the emission of several genuinely toxic pollutants (of which CO_2 is not one) into the air.

The first global cap-and-trade scheme for greenhouse gases, including CO_2, was established by the Kyoto Protocol in 1997 and finally took effect in 2005.[270] The general idea behind the protocol is to lower worldwide CO_2 production by imposing national limits or caps, in the form of tradable allowances to emit so many tonnes[271] of CO_2 per year.

These caps can be exceeded by purchasing additional allowances, known as carbon credits, either directly from other industrialized countries that emit less than their allocated amount, or through a financial exchange. In addition to trading carbon credits, industrialized nations can acquire them by investing in low-emissions projects in developing countries like China, where it costs less to cut back on CO_2 than in their own backyard. A similar option involves joint investment with lower-cost industrial countries.

Cap-and-trade provides two different methods of limiting CO_2. The first is establishing the overall cap, which can be progressively reduced in order to slow the buildup of CO_2 in the atmosphere.

The second method is the trading part: if a company emits more CO_2 than authorized by its allowances, it is penalized by having to buy more allowances or invest in development projects. But companies can profit by churning out less CO_2 than their cap amount and selling their surplus allowances to others, which provides an incentive to hold down emissions. Releases of CO_2 are governed by the market, not by heavy-handed regulations, at least in theory.

The Kyoto scheme sets legally binding targets for 38 industrialized countries, with the goal of reducing global greenhouse gas emissions by an average of 5.2% of 1990 levels, over the five years from 2008 to 2012. Several countries are allowed to increase their emissions during this period. Altogether, 181 nations had signed the protocol by early 2009.

The U.S. signed, but never ratified, the protocol because India and China – who together emit more than a third of the world's CO_2, and more than the U.S.[272] – are exempt from its requirements.

It has probably occurred to you, even after the short description above, that a cap-and-trade system is far more cumbersome administratively than a simple emissions tax on CO_2 would be. Several countries have in fact enacted a carbon tax to control CO_2 emissions. But despite growing support for carbon taxation from both environmentalists and some industrial leaders, the unpopularity of new taxes, especially in the U.S., will probably restrict the implementation of large-scale emission controls in the near future to cap-and-trade – even though this form of CO_2 regulation is really a tax in disguise.

Of course, since CO_2 has little effect on global temperatures, CO_2 cap-and-trade and carbon taxation are both a massive waste of our time and effort. Regulating CO_2 is money down the drain.

It's the economic cost of CO_2 regulation that societies and governments need to take a long, hard look at. The cost is high, because a large proportion of global economic activity depends on carbon through the burning of fossil fuels. In the U.S., around 85% of energy consumption is carbon-based, and energy use underlies nearly every sector of the economy.

What's the cost to society of lowering CO_2 emissions?

The U.S. Congressional Budget Office (CBO) – well-known for its impartial and accurate financial analyses of proposed government programs – has examined this question several times in the last 10 years, primarily for CO_2 cap-and-trade schemes. In a study conducted in 2000, based on 1998 emission levels, the CBO estimated that a 15% cut in CO_2 emissions would cost the average U.S. household from 2% to 3% of its annual income, across the whole economic spectrum.[273]

While this may not sound like much, the percentage increase is about the same as the inflation rate in the U.S. during the 2000s. It's common knowledge that those who live on a fixed income, such as seniors, quickly find themselves in a financial squeeze without annual adjustments for inflation – so a cost increase of even a few percent for a cap-and-trade system will unquestionably have a signifi-cant impact on all but the very rich.

Indeed, the CBO report points out that the burden of a CO_2 trading plan, which would raise prices for energy and energy-intensive goods and services, would fall more heavily on the poor than on the rich.[274] This is because the poor spend a larger percentage of their income on energy costs than those who are better off. So cap-and-trade is equivalent to a regressive tax.

And higher energy costs can cause bigger problems. Although rising energy prices alone are not believed to cause recessions, most economists agree that more expensive energy increases the *risk* of an economic downturn and that, along with other factors, increased energy prices will trigger a recession. As we are all well aware currently, recessions mean lost jobs as well as other misery.

Global warming alarmists argue that the cost of taking steps to limit CO_2 emis-sions far outweighs the cost of doing nothing, that the cost of regulation is a small price to pay for the benefits of not letting our climate get out of control. If there were an established link between CO_2 and global warming, I'd agree with them.

But since there's little, if any, connection, why risk another recession (or making the current one worse)?

The irony of the Kyoto Protocol for lowering global CO_2 emissions is that the protocol had its roots in the earlier UN Framework Convention on Climate Change, a treaty adopted in 1992. The treaty's stated objective was to stabilize greenhouse gas concentrations in the atmosphere in order to prevent interfer-ence with the Earth's climate, but only in ways that would not disrupt the global economy.[275] Ignoring the issue of whether the treaty was right about CO_2, the only mechanism the UN has actually been able to come up with for reducing CO_2 levels is a cap-and-trade plan that may already be contributing to the global recession.

Europe's Failed Cap-and-Trade

Twice in three years, the European Union has attempted to introduce a cap-and-trade system for CO_2. The first try was a dismal failure, and the second isn't doing much better.

This could mean simply that the Europeans don't know how to run a carbon market but, to be fair, it's much more likely that carbon caps just don't work. The first phase of the scheme was marred by administrative bungling from the outset. Yet, the second, better thought-out version is collapsing for economic reasons, even though economics should be the very underpinning of a carbon trading market.

If anything, the European experiment in cap-and-trade has achieved the opposite of what was intended. Instead of greenhouse gas emissions going down, the amount of CO_2 pumped into European skies has gone up, and the cost of emitting CO_2 has fallen through the floor. It's currently far more profitable for utilities in Europe to burn more carbon-based fossil fuels than ever before, rather than investing in carbon-free alternative energies, as was hoped.

The first CO_2 cap-and-trade market in Europe began in 2005. It included about 11,000 of the larger power plants and industrial facilities that together account for almost half of European CO_2 emissions.

The price of allowances initially rose to a peak of around €30 per metric ton ($36 per U.S. ton) of CO_2,[276] which is about the price level needed in Europe for renewable forms of energy to become competitive with electricity generated by burning coal and other fossil fuels. But by May 2006, the trading price had dropped below €10 per metric ton. Early in 2007, it had sunk to less than €1 per metric ton, reaching an all-time low under €0.10 by the end of that year.[277]

What happened essentially is that European governments, who were given the task of allocating the allowances, allocated too many, and gave most of them away free rather than auctioning them to create a more valuable commodity. Some companies, especially electrical utilities, promptly sold their excess free permits, resulting in windfall profits that were never intended as part of the plan.

In addition, national CO_2 caps, instead of being derived from the limits established under the Kyoto Protocol, were based on past emission histories for each nation. This procedure merely encouraged governments to give away large numbers of allowances to the biggest CO_2 emitters, who had little motivation to lower their emissions.

But worst of all, nearly every country sought to *increase* its base emission level over the three years that the scheme was to run, completely defeating the purpose of cap-and-trade. Only Germany and the UK asked for carbon caps lower than historic emission levels.[278]

Needless to say, verified CO_2 emissions in Europe increased between 2005 and 2007, by about 2% overall,[277] and a large number of member states did not meet their Kyoto emission targets.

These disastrous results prompted the European Union to tighten national caps for the second phase of its trading scheme, scheduled to operate from 2008 to 2012, with total allowances being cut 7% below 2005 emission levels. The allocation system for allowances was also toughened up.

But so far, the second attempt at regulating CO_2 emissions is faring as poorly as the first.

By February 2009, the price of what some traders call "hot air" had again fallen under €10 per metric ton ($12 per U.S. ton) – just as the price of CO_2 had done a year into the original cap-and-trade market, three years earlier. The impending failure of a second carbon market induced The Guardian, an English newspaper that supports the alarmist view of global warming, to declare that "Europe's whizz-bang carbon market is turning sub-prime."[279]

It's not hard to see why Europe's second cap-and-trade system is faltering, given the dire state of the global economy since trading commenced in 2008. As demand falls and factories trim their output, companies don't need as many CO_2 allowances, so they sell off the surplus. This not only brings the allowance price down, but also provides an unintended financial handout to industrial CO_2 emitters by helping them survive on fewer orders and less credit, without having to take more drastic measures to keep their businesses going.

In fact, the very introduction of CO_2 cap-and-trade in Europe may have helped tip the European economy into recession, as discussed in the previous section. Although it was the collapse of the housing market that took the U.S. economy over the brink in 2008, there were signs of rising prices for food and energy in Europe almost a year earlier.[280]

The Europeans, undaunted, are making plans for yet a third attempt at a CO_2 trading market. Admitting indirectly that the first phase failed, calling it "learning by doing", the European Union has announced a third trading period from 2013 to 2020, in which 27 national carbon caps will be replaced by a single cap for the whole Union, and many more allowances will be auctioned instead of handed out free.[281]

Nevertheless, the market's experience with the first two phases of the European scheme demonstrates clearly that CO_2 cap-and-trade is ineffective at holding CO_2 emissions in check; the third reincarnation is doomed to failure as well. Lots of euros down the drain for a measure that isn't necessary in the first place.

The U.S. Tests the Waters

Early in 2009, the Obama administration announced its plan for a U.S. cap-and-trade system, in which at least some CO_2 allowances will be auctioned as in the proposed third phase of the European Union's scheme. Congressional action on the plan was expected before the end of the year, ahead of UN discussions in December, 2009 on extending the Kyoto Protocol for greenhouse gas emissions beyond 2012.

However, the U.S. Congress had rejected earlier cap-and-trade legislation designed to reduce greenhouse gases, mostly CO_2, both in 2008 and several years before. The 2008 bill targeted electric utilities, transportation and manufacturing, setting a cap of 19% below 2005 emission levels, to be attained by 2020, and a far-reaching 71% below 2005 emissions to be achieved by 2050.[282]

The U.S. already has an emissions trading system in place for sulfur dioxide (SO_2), a toxic gas that causes acid rain when released into the atmosphere and, like CO_2, is generated from the combustion of fossil fuels.

But, while the program is often touted as a success, SO_2 emissions in the U.S. have fallen by only 25% since trading began in 1994.[283] That's no more than the decrease that occurred previously between 1980 and the program's startup – a period when there was no SO_2 regulation at all, and when power plants and other SO_2 producers took only voluntary steps to cut their emissions. So much for the effectiveness of cap-and-trade.

Despite this unimpressive performance, the SO_2 program was used as a model for the failed 2008 bill to cap CO_2 and other greenhouse gases, introduced by Senators Lieberman and Warner. Yet the principal reason for the bill's failure was not that Congress thought the plan might not work, but its enormous projected cost.

An analysis by the U.S. Environmental Protection Agency (EPA) forecast that, if the legislation were enacted, the subsequently higher cost of living would chop between \$1 trillion and \$3 trillion annually off the national economy by the year 2050, which corresponds to a reduction of 2.4% to 6.9% in U.S. output.[284] For comparison, the U.S. and European economies are expected to shrink by around 4% in 2009 during the current global recession.

One of the more immediate effects of controlling CO_2 emissions is that prices for electricity (and other forms of energy) go up, as utility companies pass on the cost of CO_2 allowances to their customers. The EPA analysis of the Lieberman/Warner bill predicted that by 2030, electricity prices would be from 35% to 79% above normal.[285]

Even higher estimates of energy prices came from the U.S. National Association of Manufacturers (NAM), who projected that households would pay between 77% and 129% more for their electricity by 2030 under the proposed legislation. Prices of natural gas and gasoline would rise by similar amounts.[286]

Further, the NAM study predicted that by 2020, higher energy prices would cost low-income families about 20% of their income, compared to 17% without greenhouse gas caps.[287] These percentages are in line with those calculated by the CBO, mentioned earlier in the chapter, and underscore the burden that cap-and-trade imposes on the poor.

The NAM also concluded that between 3 million and 4 million U.S. jobs would disappear in 2030 if the Lieberman/Warner cap-and-trade bill became law. Similar estimates have been made by other organizations.[288] The job losses would result from lower industrial output due to higher energy prices, the high cost of complying with the required emissions cuts, and greater competition from overseas manufacturers with access to cheaper energy. Energy-intensive manufacturing sectors, such as the production of steel, cement, paper and fertilizers, would suffer the most.[286]

The cost of the Obama administration's new cap-and-trade plan was originally estimated at $646 billion over eight years (from 2012 to 2019), but was later pegged at closer to $2 trillion.[268] This is essentially the same as the EPA's long-term forecast of the increased cost of living under the Lieberman/Warner scheme, but over a shorter time frame.

Even though the intention is that some of the money raised by selling CO_2 allowances will be returned to consumers in the form of lower taxes, it is generally agreed in the U.S. that the latest version of cap-and-trade will still cause economic hardship to households and businesses alike – except perhaps for utility companies who, like their European counterparts, stand to gain from the trading system.

And all for nothing, as CO_2 has little to do with global warming. Lots of dollars down the drain.

Other Countries

Several smaller countries are also looking into CO_2 regulation, notably New Zealand, Australia and Canada.

New Zealand, after briefly considering a proposed carbon tax in 2005 and then abandoning the idea, instead initiated a CO_2 cap-and-trade system in 2008. However, following a national election two months later, the new government slammed the brakes on the trading scheme, announcing that its implementation would be delayed until the scheme had been fully reviewed.[289] The move reflects differences between the two political parties in the governing coalition, one of which campaigned during the election to abolish the trading system altogether.

The review will include the present trading system, alternatives such as carbon taxation, and the consideration of "competing views on the scientific aspects of climate change from internationally respected sources".[290] So, at least New Zealanders are willing to accept the possibility that CO_2 may not cause global warming.

And, no doubt mindful of the several economic studies of proposed cap-and-trade schemes in the U.S., the review will also consider the impact of any CO_2 controls on New Zealand's economy, at both the national and the individual level, especially in light of the currently shaky state of the economy worldwide.

Specifically mentioned in the review agreement is the desire to "balance our environmental responsibilities with our economic needs".[291]

New Zealand's neighbor Australia is considering a carbon market, slated to start trading in 2011, which aims to be the world's broadest.[292] In Canada, the governing Conservative party also favors CO_2 cap-and-trade, but may not take any action until 2015. Support for an alternative carbon tax proposed by the minority Liberal party cost the party heavily in the 2008 Canadian federal election,[293] so that plan is unlikely to be revived.

Worldwide, therefore, there is no successful CO_2 cap-and-trade system in operation anywhere at this point, although the Kyoto Protocol has less than three years left to run. This is not only because of the anticipated cost of carbon caps, but also because countries such as New Zealand are beginning to question the conventional wisdom about global warming and CO_2.

The Kyoto Protocol itself has been a resounding failure. Drawn up with the intent of stabilizing global emissions of CO_2 and other greenhouse gases, the protocol set the apparently modest goal of reducing emissions by 5.2% of 1990 levels by 2012. As of 2007, worldwide CO_2 emissions had *increased* by 38% since 1992 and were still going up.[294]

And many developing countries, not bound by the treaty's emission targets to begin with, have boosted their emissions considerably – especially China, whose CO_2 output has more than doubled since 1992.

It's the developing nations, which today account for more than half of global CO_2 emissions, that will be hurt most by any future regulation of CO_2. The cost of regulation may or may not tip the economic scales for the industrialized countries, but there can be no doubt that regulatory costs will have devastating economic effects for any Third World country that institutes carbon caps or carbon taxation to lower its CO_2 output.

The overriding problem in the developing world is widespread poverty. It goes without saying that poor nations will need the cheapest sources of energy available to bring them into the industrial era and raise their standard of living.

But those energy sources won't be wind turbines or solar cells that are still too costly for extensive use, even in industrial nations. They'll be conventional power plants burning coal or natural gas, just like we employ in the U.S. to generate well over half of our electricity, because coal and natural gas are plentiful and cheap. These are the energy sources that China and India are exploiting for rapid industrialization.

Alarmists are fond of trying to argue that the developing world will have to bear the brunt of climate change, in the form of stronger hurricanes and higher sea levels that will flood low-lying nations, so that we need to limit CO_2 emissions for the sake of countries such as Bangladesh.

But as we've seen, CO_2 contributes little to global warming. And we have no control at all of the natural sources that are responsible for currently higher temperatures, such as the sun. So cutting back on CO_2 isn't going to do much to help the poorer nations on our planet cope with global warming, and will in fact be a major economic burden on them.

Lots of money down the drain – money that could be far better spent in adapting to the effects of global warming, if it continues, on developing countries.

THE FALSE PROMISE OF RENEWABLES

The big hope behind market-based regulatory schemes for CO_2 is that they will create incentives for power plants and industrial facilities to switch to alternative types of energy that don't depend on the burning of fossil fuels, and therefore don't generate CO_2. These alternatives, frequently spotlighted in the media, include nuclear power as well as renewable energy sources, such as biofuels and hydroelectric, wind, and solar power.

How realistic is this hope? Is there any chance that renewable energy, or a combination of renewables and nuclear power, could meet out future needs?

In the short term, it's highly unrealistic to think that all but a small fraction of the energy we require for electricity and transportation could come from anything but fossil fuels. This isn't just my personal opinion. It's the collective opinion of numerous experts and analysts in the energy arena, who cite the many technical barriers to be overcome in commercializing renewable energy technologies, and their present high cost.

Don't get me wrong. I'm not opposed to alternative forms of energy generation and, as a former nuclear physicist, am a huge fan of nuclear power. But to believe that we can make an appreciable dent in our use of fossil fuels such as coal and natural gas within the next 20 or 30 years is fooling ourselves. If nuclear and renewable energy are going to play a significant role, it won't be until 50 years from now.

It takes time – time to build nuclear plants, time to refine alternative technologies, time to scale them up, time to develop the infrastructure for energy distribution, and time to get the cost down to where switching over to the alternatives won't break the bank. And renewables will always be supplemental energy sources, since no single source will come close to providing sufficient power for a whole country.

A case in point is wind energy. Wind is currently favored by many environmentalists because the technology is clean, simple (basically a modern version of the old-fashioned windmill) and well-established. The principal drawback is that even in windy areas, the wind doesn't blow all the time, so wind energy can only be delivered intermittently.

Nevertheless, the U.S. Department of Energy (DOE), in conjunction with wind turbine manufacturers, utility companies and others, has recently studied the feasibility of supplying 20% of the nation's electricity from wind energy by 2030.[295]

That's a tall order, given that the contribution of wind to U.S. electricity demand was only 1.2% in 2008. Several European countries produce substantial quantities of wind power, led by Denmark where 19% of the electricity generated comes from wind – but this amount represents only 0.2% of electricity consumption in the U.S.,[296] as Denmark is so small.

The study concedes that a 20% wind scenario by 2030 is ambitious and would require "significant changes in transmission, manufacturing, and markets".[295] The largest challenge will be in transmission, since wind energy can't be stored and most of it will be produced in sparsely populated areas of the country, which are also the windiest. The DOE study estimates that construction of the necessary transmission lines will add about 10% to the capital cost of building the extra wind energy capacity, a cost that will be borne by consumers.

But neither technical challenges nor cost are the main issue here. If the purpose of manufacturing enough wind turbines to meet 20% of the U.S. demand for electricity is to get away from burning fossil fuels, then the exercise is another massive boondoggle.

By the study authors' own admission, generating that amount of electricity from wind in 2030 will avoid only about 10% of the CO_2 emissions that would

TABLE 8.1: CARBON CAPS

Control of CO_2 emissions

- The UN Kyoto Protocol mandates the lowering of CO_2 emissions from 38 industrialized countries, by an average of 5.2% of 1990 levels, before 2012.

- To achieve this target, the European Union has introduced a CO_2 cap-and-trade system. Trading of the initially free CO_2 allowances began in 2005.

- The U.S. will probably adopt a similar cap-and-trade scheme in 2009 or 2010, except that some CO_2 allowances will be auctioned.

- Several other industrial nations are also considering carbon caps.

- Cap-and-trade systems for CO_2 are expected to speed up development of renewable forms of energy that don't generate CO_2, such as wind power.

WHAT'S WRONG WITH CARBON CAPS?

1. Controlling CO_2 emissions will have little impact on global warming, which most likely comes from natural causes.

2. Cutting back on CO_2 will cost trillions of dollars – wasted money that could be used instead for adapting to any future effects of global warming.

3. Europe's first cap-and-trade system failed, with the price of CO_2 allowances reaching rock bottom in 2007 while European emissions increased. The second phase, with tighter CO_2 caps, isn't doing much better so far.

4. Cap-and-trade increases the cost of living by raising energy prices, which can in turn nudge the economy into recession.

5. The cost of carbon caps falls most heavily on the poor. This cost will be a major economic burden for the developing countries, as well as the less well-off in industrialized nations.

6. Switching to renewable energies won't reduce overall CO_2 emissions, both because renewables alone will be unable to meet total electricity demand, and because the unpredictability of wind and solar power will require backup energy from fossil fuels.

7. Cap-and-trade schemes cause job losses that outnumber the green jobs created in renewables.

occur without any significant boost in wind power.[297] Projected CO_2 emissions from all energy sources in 2030, even with all that additional wind energy, will still exceed today's emissions by 7%, from DOE statistics.[298]

And according to a Canadian report, experience in Denmark – the world's most wind-intensive nation, with more than 6,000 turbines generating 19% of its electricity – has been that additional coal-generated electricity is needed to cover the unpredictability of wind power.[299] That has led to rising Danish CO_2 emissions, by 16% in 2006 alone.[300] Devotees of the Kyoto Protocol can't be too happy.

I'm all in favor of developing renewable energies such as wind, in the U.S. and elsewhere. But let's not delude ourselves that taking this path is going to reduce global CO_2 emissions or have any noticeable effect on global warming.

In any case, not everyone is as optimistic about the potential for renewables as the authors of the wind scenario study. The DOE's annual report for 2009 predicts that only 2.5% of U.S. electricity generation in 2030 will come from wind power, not the 20% projected in the study, and that the total contribution from *all* renewable sources will be only 14%.[301] In fact, the largest percentage of renewable energy is expected to come from biofuels that, like fossil fuels, produce CO_2 when burned – though at a reduced level.

In 2030, coal and natural gas combined will still be providing two thirds of U.S. electricity, down only slightly from today,[302] if we don't want to go back to living in caves. Even the wind study recognizes that "coal power will continue to play a major role in future electricity generation".[295]

So we'll still be putting lots of CO_2 into the atmosphere unless, perhaps, we can develop so-called clean coal technology. This entails capturing the CO_2 generated by burning coal and other fossil fuels, and storing it deep underground. But the technology is unproven and is likely to be extremely expensive.

Solar power was once a leading contender for the energy source of the future. However, while solar technologies play an important role in sunny, remote parts of the globe where there is no electric grid, and on satellites, they are still much too costly for grid-connected use in the developed world. Like wind energy, solar energy can't be easily stored and, of course, the sun doesn't shine at night.

The simple truth is that renewables don't, and won't in the future, have the capacity to meet the world's thirst for energy. Currently, the largest source of renewable energy in the U.S. is hydroelectric power, but this won't grow beyond its present level because the country has run out of unused reservoirs in the mountains.

An argument often used to justify the anticipated high cost of switching to renewable forms of energy is that they will create jobs – green jobs. One of the major newspapers in the U.S. Midwest, home to much of the country's heavy manufacturing, recently carried a full-page advertisement from an environmental group, claiming that carbon caps will get out-of-work steelworkers back on the job.

Cleverly designed for emotional appeal, the ad showed five flannel-shirted workers laid off from their steel mill, with a tag line suggesting that carbon caps will reopen the mill so the workers can make steel for wind turbines. Never mind that cap-and-trade will cost jobs – especially in the steel industry, as mentioned before – nor that it will raise energy prices and the cost of living.

Whoever designed the carbon cap advertisement can't do arithmetic.

The DOE wind scenario study calculates how many jobs would be created by building the turbines needed to achieve the study's goal of 20% wind energy capacity by 2030, using a standard economic impact model. The results wouldn't excite too many economic development departments, with an average of 260,000 new jobs created annually between now and 2030 in manufacturing, construction, turbine operations and related businesses. That's for the whole U.S. In most states, there would be under 10,000 new jobs per year.[295]

Even in the peak years from 2020 to 2030, the annual job gain from building wind turbines would be only about 500,000 nationally. Including the new jobs in other renewable energies could lift this number to, at most, around 1 million annually during that time. Compare that to NAM's estimate of between 1.5 million and 3.5 million jobs lost each year for the same period,[286] as a result of the CO_2 cap-and-trade bill debated by the U.S. Congress in 2008.

These numbers are all projections. But there are real numbers based on actual experience available from Spain, which is part of the European Union's cap-and-trade system. A Spanish study found that for every new job created in renewables, particularly wind energy, slightly more than two jobs are lost. Added the study's author: "The loss of jobs could be greater if you account for the amount of lost industry that moves out of the country due to higher energy prices."[303]

Creating green jobs is an admirable objective, but not if it destroys existing jobs in traditional industries. Carbon caps simply don't add up.

Futhermore, the negative economic effect of job losses from carbon caps is compounded by the heavy subsidization of new jobs in renewable energies. According to the DOE, U.S. government subsidies in 2007 amounted to $23.37 per megawatt hour for electricity produced from wind, compared to only 44 cents for coal and $1.59 for nuclear power.[304] Although the subsidies for renewables will eventually come down, the present subsidy cost will be passed along to consumers.

Subsidies, insufficient capacity, and job losses – that's what we can expect for renewable energy, if we force the pace of development as cap-and-trade will. And while renewables will slow the growth of CO_2 emissions slightly, there's no sign that the CO_2 level will actually fall. Of course, it doesn't matter much if the CO_2 level falls or not anyway, as far as global warming is concerned.

Chapter 9: Summary

An Internet joke popular with skeptics asserts that global warming comes from a decrease in the number of pirates today, compared to the 19th century,[305] a joke recently reinforced by the upsurge in Somali piracy that has accompanied declining temperatures.

It makes about as much sense to attribute global warming to diminishing piracy as it does to connect it with man-made CO_2 in the atmosphere. There's precious little evidence that the higher temperatures of the last 160 years are caused by anything but natural climate variability.

So why do global warming alarmists, taking their cue from the IPCC, believe otherwise? Exploring the psychological answers to this question would take another book. I'll confine myself here to summarizing the arguments made in the present book against the case for human-induced global warming – a case that is embodied in what I've called the CO_2 hypothesis and supposedly upheld by the results of computer climate simulations.

The CO_2 hypothesis was first put forward by Sweden's Svante Arrhenius in the early 1900s, though it wasn't taken seriously until the end of the century when the IPCC began publishing its climate reports. The hypothesis attributes global warming to greenhouse gases, primarily CO_2 that comes from burning fossil fuels.

More than 200 years after industrialization began pumping CO_2 into the Earth's atmosphere, the CO_2 level is still low: about 390 parts per million. Alarmists say that's enough to wreak havoc with our climate. But we need to understand that such a small quantity of CO_2 can't raise the mercury very much by itself, just from the greenhouse effect.

According to theory, global temperatures should increase by only 1.2° Celsius (2.2° Fahrenheit) when the CO_2 concentration doubles (see Table 5.2), and we're nowhere near that point yet. The only way that temperatures could have gone up to where they are now, if the CO_2 hypothesis is valid, is through an amplification process. That requires the tiny effect from CO_2 alone to be given a boost by water vapor in the atmosphere and by clouds.

The technical term for this amplification is feedback, which I discussed in Chapter 5. Positive feedback or magnification enhances the climate's sensitivity to minute amounts of added CO_2, producing a larger temperature rise than expected. Negative feedback or diminution lowers the climate sensitivity, and reduces the warming from CO_2 to a negligible level.

One of three big red flags for the CO_2 hypothesis is that recent satellite observations show cloud feedback to be negative – which means that clouds reduce global warming – rather than positive, as the IPCC's computer models predict. If this negative cloud feedback is strong enough to overcome the predicted positive feedback from atmospheric water vapor, or if water vapor feedback also turns out to be negative, then overall CO_2 feedback must be negative and CO_2 simply can't be the cause of global warming.

The other two red flags involve matching up CO_2 and temperature records for the last 2,000 years. If the IPCC and alarmists are right about CO_2, past temperatures should mimic the CO_2 level over long periods of time, since CO_2 is postulated to be the source of global warming.

Trouble is, the temperature doesn't track the CO_2 level for thousands of years, during a period starting around the end of the last ice age and lasting until 160 years ago. The temperature record does synchronize with the CO_2 record during the ice ages, as well as over the last 160 years (this being the origin of the global warming scare). But for most of the intervening interval, the temperature was all over the map during a time when the CO_2 concentration held rock steady.

This mismatch between CO_2 and temperature didn't go unnoticed by the IPCC. Desperate to have the CO_2 hypothesis accepted, the IPCC stooped to trickery and rewrote history to make the temperature and CO_2 records conform during that inconvenient period, creating the so-called hockey stick graph that wiped out the well-documented Medieval Warm Period and Little Ice Age.

Fortunately for the good name of science, the hockey stick was completely discredited, though not until several years after it was first published by the IPCC.

However, this leaves the IPCC and global warming alarmists with the problem of how to explain the disparity between temperature and CO_2 during the Middle Ages. That they can't explain it is a nail in the coffin of man-made global warming.

The third red flag for the CO_2 hypothesis is more subtle, involving the CO_2 level lagging behind temperature in the ice age era. While the lag can be largely explained by the delayed release of stored CO_2 from the world's oceans, the lag means that during global warming at the termination of an ice age, atmospheric CO_2 kept rising for 600 to 800 years after the temperature had flattened out.

The dilemma in this observation for advocates of the CO_2 theory of modern global warming is that since 1850, the CO_2 level and temperature have increased almost simultaneously. But either temperature and CO_2 go up and down at the same time, or they don't. You can't have it one way during the ice ages and another way today. This inconsistency is yet another nail in the man-made warming coffin.

But the IPCC and its alarmist followers don't buy any of this. There can't be *any* red flags. The only way that global warming alarmists can patch together a coherent climate picture that implicates CO_2 is to deny sound scientific evidence. Although alarmists like to disparage skeptics as deniers, it is predominantly the alarmists who suppress data and twist the facts to suit their own agenda, which is to leave no doubt in the public mind that global warming comes from human activity.

To carry out the alarmist agenda, the IPCC uses several weapons of mass deception.

Its weapon of choice is data manipulation – bending the truth to make the data conform with the CO_2 hypothesis at any cost. The most glaring example of this is the hockey stick episode, involving the historical temperature record. But the IPCC has meddled in other areas as well.

These include ignoring bias in the modern temperature record, caused by urbanization, that inflates global warming; ignoring the shutting down of weather stations in cold parts of the world during the 1990s, that artificially increases the recent warming rate; cherry-picking 19th century CO_2 data, so as to exaggerate the rise in CO_2 level since preindustrial times; and trivializing the sun's contribution to the present warming trend.

Sometimes the IPCC goes even further, as it did in trying to squelch statistical evidence for temperature bias, by utilizing another weapon: data fabrication. This has led to an accusation by one of the IPCC's own reviewers that the panel made false claims and cooked up evidence, and to allegations of data tampering by a manager of solar satellite data. In yet another instance, actual charges of fraud have been leveled against an IPCC climate scientist.

But what the IPCC falls back on above all else, to shore up its dubious assertion that man-made CO_2 causes global warming, are its computer climate models.

It's a sign of the times, and of today's embrace of junk science, that the projections of these computer models are rarely questioned, even though the models are full of unfounded assumptions. One of the biggest assumptions in most models is that the sensitivity of the climate to CO_2 is just the same today as it was during the ice ages, even though the climate was very different back then.

And there's all the difference in the world between a computer model used to design, say, an airplane and a computer climate simulation. Most of us have few qualms about traveling through the skies in a pressurized tube that sprouts wings, because we know somewhere – at least at the back of our minds, if we're not engineers or scientists – that our flying machine obeys the well-known laws of aerodynamics. So much so that we're completely comfortable when the pilot switches over to autopilot and the plane basically flies itself.

The Earths' climate system is a far cry from a jet plane. I don't think it's too much of a stretch to say that our understanding of climate is still in its infancy. It's very presumptuous for us to believe we're anywhere near unraveling the many mysteries of climatology, especially the role played by greenhouse gases.

We know something about the sun and the atmosphere, about oceans and wind, but there's a whole lot we don't know about clouds, water vapor, atmospheric wind patterns, ocean heat and salinity, and cosmic rays – not to mention a host of other climatic variables and influences. All of these are incorporated in computer climate models as adjustable parameters, generally numbering in the hundreds.

Scientists in other fields sometimes agonize over the precise value of just a single adjustable parameter, often to a high degree of accuracy. Why should we pay any attention to the results of computer models containing multiple parameters, when most of the parameters can't be specified very accurately at all? No wonder so many climate predictions made by the IPCC's models have turned out to be wrong.

Even the most fundamental predictions about temperature made by the IPCC are incorrect. All its models project ever increasing global warming, from the year 2000 through the end of the century. Yet temperatures have been falling since 2002.

I'm not saying the mercury can't start going up again, but IPCC computer models never even hinted that it might go down. And it's not just surface temperatures that have decreased: the ocean depths have stopped warming as well. Also wrong in the climate models are the temperature trends at the North and South Poles, where global heating should be strongest but is barely evident at all, and the warming rates at different levels of the atmosphere that are exactly the opposite of what the IPCC predicts. Forecasts of stronger hurricanes and of open water at the North Pole instead of an ice cap haven't been borne out.

Could the models be any worse?

In essence, IPCC computer climate models are preprogrammed to deliver the result that a minuscule amount of CO_2 in the atmosphere is the dominant source of global warming.

Everything that conflicts with this result is either left out of the simulations altogether, or minimized – by deceit if necessary. This highly selective and scientifically dishonest approach is behind the IPCC claim that its computer models can only reproduce the 20th-century temperature record if they omit all sources of natural variability except the sun, which the models barely account for in any case.

Never mind that there is abundant evidence that the sun's role in global warming has been grossly underestimated in computer climate models, nor that the models don't even include the 1,500-year climate cycle that is probably tied to the sun, nor that they ignore indirect warming from solar blocking of cosmic rays.

And never mind the various natural climate cycles that the IPCC can't even model adequately, let alone consider as possible alternatives to CO_2 as the source of global warming. A recent proposal based on satellite data links global warming to cloud fluctuations associated with the Pacific Decadal Oscillation, a natural cycle that switches between warm and cool phases every 30 years, and that shows a striking similarity to the temperature record since 1850. Needless to say, the IPCC's models simply assume that such natural variability doesn't even exist, and that the only possible explanation for climate change is greenhouse warming.

Deception, data manipulation, data fabrication, fraud, unbridled faith in deficient computer models – that's quite a string of indictments against the IPCC. The IPCC has perverted science and tarnished the age-old scientific method, all to prop up the baseless assertion that man-made CO_2 causes global warming. If CO_2 plays any role at all, it's a very small one.

Yet the IPCC insists that it can say with 90% confidence that human activities since 1750 are the source of global warming, and that it is 90% sure of even higher temperatures in this century if we continue to emit CO_2 and other greenhouse gases. It's these high IPCC confidence levels, totally unjustified by the evidence, that have led to the false and scientifically untenable belief that a consensus exists on human-induced global warming.

Human nature being what it is, not all skeptics are paragons of scientific virtue either. I've come across Internet blogs in which authors bend the truth in defense of the skeptical cause – though such distortion is completely unnecessary even from a political viewpoint, with ample evidence to support the case against CO_2 as we've seen.

Were that the end of the story, there wouldn't be too much cause for concern. Science has undergone comparable attacks in the past and has survived, with corruption and suppression of the truth eventually yielding to renewed belief in the merits of the scientific process.

But the IPCC has spun its web of deceit so far and wide that, by most counts, a majority of the public adheres to the IPCC view that CO_2 emissions must be drastically curbed for the good of the planet, and the sooner the better. The problem with this erroneous alarmist message is that action on CO_2 is going to cost the world trillions of dollars, to fix something that doesn't need to be fixed in the first place. That's what it will cost to regulate CO_2 worldwide, money that could be better spent dealing with the effects of global warming, if it resumes.

Nonetheless, the industrialized nations have either embarked on the CO_2 regulatory road already, or are about to, and the rest of the world will be expected to follow. After all, you can't have half the world reducing their emissions and the other half doing nothing, if you truly believe globally distributed CO_2 is the problem. But apart from the enormous expense, achieving such a goal won't be easy.

Europe is four years into a cap-and-trade system for CO_2, the first phase of which was a dismal failure. Although the Europeans admit they made mistakes in implementing the scheme, the second phase isn't faring any better. The European Union's CO_2 emissions continue to go up, and the price of CO_2 allowances is in the doldrums.

Numerous studies have shown that cap-and-trade hurts the economy by raising energy prices. Utility companies simply pass on to their customers the costs of the CO_2 permits they are required to purchase.

As a result, most of the financial burden of this completely misguided global effort to control CO_2 emissions falls on the poor. The cost of carbon caps will be especially devastating for developing countries, already struggling to catch up to their industrial neighbors.

The alarmist response to such social concerns is that carbon caps are needed to prod energy companies into developing alternative, CO_2-free sources of energy, such as nuclear power and renewables. It may take a costly cap-and-trade scheme, or some form of carbon tax, to do just that. But to think that renewable sources can provide more than a fraction of our future energy needs, or that turning to renewables is going to cut CO_2 emissions very much, if at all – or is going to create jobs – is an illusion.

A prevalent belief among environmentalists is that wind turbines and solar cells can meet the bulk of the world's electricity demand. But as Earth-friendly as this belief may be, the reality is that neither the wind nor the sun can deliver energy on a constant basis, nor can the energy be readily stored. So steadier, backup sources are required, and that means continuing to rely on fossil fuels such as coal, as recent Danish experience has shown.

But all this is a massive exercise in futility anyway, because CO_2 has little to do with global warming. If we could turn our attention away from CO_2 to one of humanity's real troubles, future generations are much more likely to thank us than if we squander our resources on an imaginary problem.

You may be wondering just how we came to be in this position, how so many prominent figures and so many learned societies, and the mass media, have sided with the IPCC on the assumed connection between CO_2 and global warming – when this connection doesn't stand up to close scrutiny. To explore that question would take another book.

However, it's not hard to understand how so many people became duped, if you glance again at the modern temperature record in Figure 7.2 (or Figure 2.1). What you notice on viewing the chart through half-closed eyes, so that all you see is the trend line, is that the temperature really shot up over the last 40 years. Or appears to have done, for we know now that the rise is exaggerated by the urban heat island effect. This apparently steep rise since 1970 is what started the panic, with the IPCC fanning the flames by linking the warmth to climbing CO_2.

But like a roller coaster ride or a stock price trend, the temperature ascent eventually came to an end. Currently, the planet is in a cooling mode, just as it was twice before during the period covered by the chart. Yet the CO_2 level has been going up relentlessly during all those years, regardless of whether the thermometer was rising or falling at the time.

There is an *underlying* upward trend in temperature, but that simply reflects the end of the Little Ice Age around 1850. Global temperatures, like so many aspects of nature, go through cycles – and cycles sometimes result in global warming.

The world has become carried away by a short-term trend, which is not actually not so unusual in the history of science. Scientific researchers often veer off temporarily on the wrong track. But they usually keep quiet until they're certain they're back on the right track. The IPCC has chosen to trumpet its half-baked conclusions on CO_2 and global warming as the final word on the subject.

Appendix A:

Climate Feedbacks and Sensitivity

Climate sensitivity can be expressed either as the temperature change caused by a doubling of the atmospheric CO_2 level, usually from its preindustrial level in 1850, or as a measure of the climate feedbacks that contribute to that sensitivity.

It is common to define a climate feedback parameter (measured in units of W/m^2 per °C), which is the inverse of the climate sensitivity parameter (measured in units of °C per W/m^2).[306] The IPCC has created some confusion over the use of these terms by employing the same symbol λ for both of them at different times: its 2007 Fourth Assessment Report refers to λ as the climate feedback parameter,[307] while the 2001 Third Assessment Report defined λ as the climate sensitivity parameter.[308] But the switch in terminology does bring the IPCC in line with current usage by most climatologists.

The climate feedback and sensitivity parameters are defined by:

$$\text{Feedback parameter } \lambda = \Delta F / \Delta T, \qquad (A1)$$

$$\text{Sensitivity parameter} = 1 / \lambda = \Delta T / \Delta F, \qquad (A2)$$

where ΔF (in units of W/m^2) is the external radiative forcing, generally corrected for the heat flux into the oceans, and ΔT (in °C) is the change in global surface temperature due to that forcing. From Equation (A1) or (A2), the climate sensitivity ΔT is then:

$$\text{Sensitivity } \Delta T = \Delta F / \lambda, \qquad (A3)$$

where λ is the feedback parameter.

For greenhouse gases such as CO_2, well-mixed into the atmosphere, a good approximation to the radiative forcing is:

$$\Delta F = 5.35 \ln(C/C_0),[309] \qquad \text{(A4)}$$

in which C_0 and C are the concentrations of CO_2, before and after the forcing ΔF takes effect. C_0 and C are usually measured in units of parts per million by volume (ppmv).

The value of ΔT corresponding to doubled CO_2 ($C = 2C_0$) is the most commonly used measure of climate sensitivity, sometimes known as the equilibrium climate sensitivity, but the sensitivity can be calculated for any change in CO_2 concentration.

While the climate sensitivity should in principle include other greenhouse gases such as methane, it turns out that the forcing from CO_2 alone is almost the same as the net forcing from all human sources included in IPCC climate models – which embrace both positive forcings such as greenhouse gases and negative forcings such as aerosols.[310] Therefore, the sensitivity for CO_2 is a good measure of the overall climate sensitivity.

Values of the feedback parameter and the climate sensitivity, either taken from the technical literature or calculated from the equations above, are shown in Table A.1 for a number of different climate models. The climate sensitivity numbers were presented previously in Table 5.2.

Table A.1: Calculated CO_2 Climate Sensitivity

Model	Feedback parameter[a]	Predicted temperature change Today	Predicted temperature change At doubled CO_2
IPCC (2001)	0.9-2.1[311]		3.5°C[312]
IPCC (2007)	0.7-2.0[313]	0.76°C[314]	3.3°C[315]
Hansen[316]	1.4	0.6°C	2.7°C
Spencer[317]	up to 8.3	0.21°C[b]	0.45°C[c]
Zero CO_2 feedback	3.2[318]	0.54°C[b]	1.2°C[c]
Positive CO_2 feedback	< 3.2	> 0.54°C	> 1.2°C
Negative CO_2 feedback	> 3.2	< 0.54°C	< 1.2°C

[a] Measured in units of W/m² per °C
[b] Calculated from Equations (A3) and (A4), with C = 387 ppm (its 2009 level), and C_0 = 280 ppm (the preindustrial level, according to the IPCC[319])
[c] Calculated from Equations (A3) and (A4), with C = $2C_0$

The IPCC models, which are mostly so-called Atmosphere-Ocean General Circulation Models (AOGCMs), are based on computer simulations of the Earth's climate, the climate sensitivity often being determined by matching the model output to paleoclimatic conditions. For its 2001 and 2007 reports, the IPCC used as many as 34 and 23 different models, respectively.[320, 321] James Hansen's model, which is one of the current IPCC models and was developed at the Goddard Institute for Space Studies (GISS), is listed separately.

Roy Spencer's model is a much simpler climate model based on energy balance, and has been used to interpret recent satellite observations of the Pacific Decadal Oscillation (PDO) in terms of negative cloud feedback. This interpretation is discussed in Chapter 6 and Appendix B.

The range of feedback parameters shown for the IPCC models in Table A.1 reflects the variation between the different models. The climate sensitivities for doubled CO_2, however, are means over a large number of the models; the feedback parameters that correspond to the calculated sensitivities of 3.5°C (2001) and 3.3°C (2007) are both close to 1.1 W/m² per °C. Expressed in terms of the climate sensitivity parameter, which relates the temperature change to its associated forcing, this is a sensitivity of 0.9°C per W/m².

The climate sensitivities from IPCC models that have been matched to ice-age climate data range from 2.3°C to 3.7°C,[322] corresponding to feedback parameters ranging from 1.6 to 1.0 W/m² per °C, respectively. Higher feedback parameters signify lower climate sensitivity, as can be seen from Equation (A3).

In Tables A.1 and 5.2, the 2007 IPCC value of the sensitivity for the present climate is 0.76°C, exactly the same as the IPCC's stated temperature increase since 1850,[323] because the computer simulation results are fitted to this increase. The present-day climate sensitivity of 0.6°C from the Hansen model is lower because it (as well as the value for doubled CO_2) is calculated relative to 1880, rather than 1850 when global temperatures were slightly lower (see Figure 2.1).

With the exception of the 2007 IPCC number, the present-day climate sensitivities in the two tables depend on the preindustrial baseline level C_0 that is assumed for the atmospheric CO_2 concentration. If this is higher than the IPCC says, as discussed in Chapter 3, the sensitivity values are lower than those shown. Today's climate sensitivity for zero CO_2 feedback, for example, falls from 0.54°C to 0.32°C if C_0 = 320 ppm instead of 280 ppm. However, from Equations (A3) and (A4), the sensitivities for doubled CO_2 ($C = 2C_0$) are the same regardless of the baseline level, the only change for a higher baseline being a higher final CO_2 level.

Although Tables A.1 and 5.2 don't indicate any errors in the estimates of feedback parameter or climate sensitivity, which is for simplicity of presentation, all the numbers shown are subject to uncertainty. For example, the IPCC's 2007 estimate of 3.3°C for the mean climate sensitivity is actually 3.29±0.69°C, where the error is given as ±1 standard deviation.[315]

Feedback and amplification

As described in Chapter 5, positive feedbacks amplify the Earth's natural greenhouse effect for CO_2, while negative feedbacks diminish the effect.

Without any feedback at all, the climate sensitivity for doubled CO_2 is 1.2°C, corresponding to a feedback parameter of 3.2 W/m² per °C (Table A.1). This value of the feedback parameter, sometimes called the Planck feedback parameter, represents the infrared energy (3.2 W/m²) that the Earth would radiate away in response to a sudden warming of 1°C, as calculated by computer climate models.[318] The calculation is based on the Stefan-Boltzmann law of blackbody emission, which governs the temperature dependence of emitted longwave (infrared) radiation.

In the absence of other climate feedbacks, radiative temperature damping by the Earth is a strongly negative natural feedback.

Sensitivities above 1.2°C, which correspond to feedback parameters less than 3.2 W/m² per °C, are associated with net positive CO_2 feedback; sensitivities below 1.2°C, corresponding to feedback parameters greater than 3.2 W/m² per °C, imply that the net CO_2 feedback is negative.

Amplification factors for both types of feedback are shown in Table A.2. The amplification (or gain, in electronic terminology) is defined as

$$\text{Amplification} = \Delta T / \Delta T_0, \tag{A5}$$

where ΔT is the climate sensitivity with feedback and ΔT_0 is the sensitivity without feedback.

Table A.2: CO_2 Amplification Factors

Type of Feedback	Model	Feedback parameter[a]	Amplification[b]
Positive	IPCC (2007)	2.1[c]	2.9 (190%)
Positive	Hansen	1.8	2.3 (130%)
Zero	None	0.0	1.0 (0%)
Negative	Spencer	down to -5.1	0.4 (-60%)

[a] Measured in W/m² per °C, relative to $\lambda_0 = 3.2$ W/m² per °C
[b] At doubled CO_2
[c] Corresponding to a value of 1.1 W/m² per °C for the IPCC (2007) feedback parameter in Table A.1

The amplification can be calculated directly from the climate sensitivity numbers in Table A.1, using Equation (A5), or from the feedback numbers in Table A.1, using the equation:

$$\text{Amplification} = \frac{1}{1 - \lambda/\lambda_0},^{324} \tag{A6}$$

in which λ is the feedback parameter measured relative to the zero feedback value of $\lambda_0 = 3.2$ W/m² per °C, and $\lambda = \lambda_1 + \lambda_2 + ... \lambda_n$, where n is the number of independent feedbacks. Under the sign convention for Equation (A6), positive or negative feedback parameters have a positive or negative sign, respectively, and λ_0 is taken to be positive.

The amplification expressed as a percentage is also positive or negative for positive or negative feedback, respectively. Positive feedback or amplification results in magnification of the climate system's temperature response to CO_2 without feedback, while negative feedback or amplification results in diminution of the response.

When the feedback parameter in Table A.1 approaches zero, which means the relative feedback parameter in Table A.2 approaches λ_0, the amplification and the climate sensitivity become infinite (Equations (A3) and (A6)). This corresponds to so much positive feedback that a runaway greenhouse effect occurs, which could lead to a hothouse climate on Earth. Yet even the IPCC does not suggest that's likely.

Finally, we should note that everything on climate feedbacks and sensitivity summarized here is based on the assumption that the feedbacks are linear – that is, CO_2 and other feedbacks are linearly dependent on the temperature response. But the Earth's climate system is highly *nonlinear* in many ways, and can even be considered chaotic (showing apparently random behavior) in a mathematical sense. So any calculations of feedback and climate sensitivity should in principle be founded on nonlinear feedback theory.

Inclusion of nonlinear quadratic terms in the analysis of feedbacks from IPCC climate models appears to show significant nonlinearities, especially in the feedbacks from high clouds and temperature/altitude (lapse rate).[325] This underlines how little is really understood about climate feedbacks.

Appendix B:

The PDO Cloud Fluctuation Model

As discussed in Chapter 6, Roy Spencer and his research team at the University of Alabama in Huntsville have proposed a model to explain global warming, based on cloud fluctuations associated with the Pacific Decadal Oscillation (PDO).[326] The PDO is a natural cycle of climate variability that repeats approximately every 60 years, with a 30-year warm phase followed by a 30-year cool phase.

Figure B.1 shows how global temperatures have followed the pattern of the PDO over the last century.

Most of the features in the 20th century temperature record can be explained by Spencer's energy balance model, which assumes variations in cloud cover that are proportional to the PDO index, and a slab ocean that mixes heat on a time scale ranging from decades to centuries.

Evidence linking fluctuations in globally averaged cloudiness to the PDO was found in satellite observations of the Earth, made over a five-year period from 2000 to 2005. Spencer compared recent variations in the PDO index to measurements of emitted longwave (LW) infrared and reflected shortwave (SW) solar radiation over the global oceans (Figure B.2).[329]

Figure B.1: 20th Century Temperatures and the PDO[327]

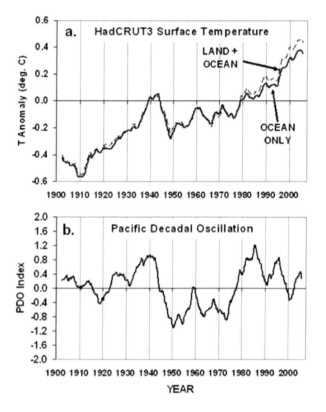

Source: Spencer.[326] The thermometer temperature record in (a), which is similar to Figure 2.1, is from the Climatic Research Unit (CRU).[328]

Because the measured radiative flux is a combination of forcing *and* feedback, Spencer used a method developed by his team[330] to isolate any time-varying radiative forcing, potentially associated with the PDO, from the feedback component. This method is based on the fact that the temperature response always lags its related forcing – by 90 degrees for harmonic forcing. What that means here is that the radiative forcing due to PDO cloud fluctuations completely obscures the feedback, which must first be removed from the data in order to see the forcing.

By plotting the observed radiative fluxes against satellite temperature data from the lower troposphere, for the period from 2000 to 2005, the feedback parameter was estimated at 8.3 W/m² per °C. This represents strongly negative feedback (see Table A.1) but, as Spencer points out, it does not necessarily correspond to long-term climate sensitivity – rather, it represents the feedback over the short term, feedback that was removed to isolate the forcing signal.

Figure B.2: The PDO Index and Ocean Radiative Fluxes

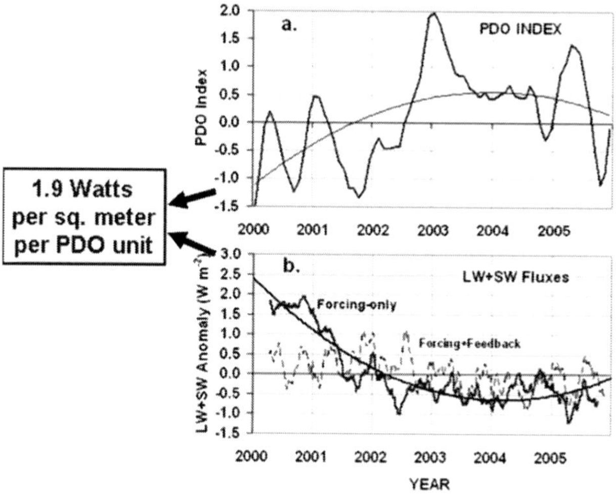

Source: Spencer.[326] The PDO index in (a) is a three-month running average; the satellite data for the radiative fluxes in (b) were analyzed to remove the feedback component. The smooth curves are 2nd order polynomial fits to the data.

With the feedback signal taken out, the smoothed PDO index in Figure B.2 is seen to be a good match to the smoothed radiative forcing, which the model assumes is due to cloud fluctuations associated with the PDO.

The value of the radiative forcing was found to be 1.9 W/m^2 per PDO index unit. This means that for each unit of the PDO index, the warm phase of the PDO was accompanied by 1.9 W/m^2 of extra heating, and the cool phase of the PDO by that much cooling.

Spencer adds that in this analysis, the PDO is not an index of temperature, but instead an index of radiative forcing, which drives the time rate of change of temperature. So the PDO index is directly related to the change in temperature with time, not to the temperature itself.

Glossary

cap-and-trade: a market trading scheme that limits emissions of a toxic pollutant, or of a greenhouse gas such as CO_2, into the atmosphere

carbon cap: the limit imposed on annual CO_2 emissions in a cap-and- trade scheme

CBO: U.S. Congressional Budget Office

Celsius: the metric temperature scale

climate sensitivity: the response of the Earth's climate system to a forcing such as solar radiation or added CO_2 in the atmosphere; the sensitivity to CO_2 is often measured as the temperature increase caused by a doubling of the CO_2 level

CO_2: carbon dioxide

DOE: U.S. Department of Energy

El Niño: a natural climate cycle that causes temperature fluctuations and other climatic effects in tropical regions of the eastern Pacific Ocean; sometimes known as the El Niño – Southern Oscillation (ENSO)

EPA: U.S. Environmental Protection Agency

Fahrenheit: the U.S. temperature scale

feedback: a response to a disturbance (forcing) that feeds back to modify the disturbance itself, either magnifying or diminishing it (positive or negative feedback, respectively); "zero" feedback in climatology corresponds to no change in the heat energy normally radiated away by the Earth – a situation that electronics engineers would describe as negative feedback, since heat is still being lost

forcing: radiative forcing, which is a disturbance that alters the climate system and that usually gives rise to feedback processes; related to climate sensitivity

fossil fuel: a fuel such as coal, oil or natural gas that is produced over millions of years by the decomposition of buried fossils, and that gives off CO_2 when burned

greenhouse gas: a gas in the atmosphere such as water vapor, CO_2, methane or nitrous oxide, all of which can trap heat radiated into space by the Earth; the main greenhouse gas is water vapor

hockey stick: the name given to the erroneous IPCC graph published in 2001, showing reconstructed temperatures for the past 1,000 years and resembling a hockey stick on its side

ice age: an extended period of severe global cooling in the past, lasting for tens or hundreds of thousands of years, when much of the Earth (especially the Northern Hemisphere) was covered by vast ice sheets and glaciers

ice core: a cylindrical core of ice extracted from an ice sheet by hollow drilling; air bubbles trapped in the ice provide an historical proxy record of past temperatures and CO_2 levels

IPCC: the UN's Intergovernmental Panel on Climate Change

Kyoto Protocol: a UN protocol to limit emissions of CO_2 and other greenhouse gases, through a cap-and-trade system that mandates carbon caps for 38 industrialized countries; the protocol, which has not been ratified by the U.S., took effect in 2005

La Niña: the cool phase of the El Niño – Southern Oscillation (ENSO), often following the El Niño warm phase

Little Ice Age: an unusually cool period, but not as cold as the glacial ice ages in the Earth's distant past, that lasted from about 1500 to the beginning of modern global warming around 1850

Maunder Minimum: the period from 1645 to 1715, when the sun's activity was low and the annual number of sunspots was close to zero

Medieval Warm Period: the previous period of global warming, lasting from approximately 800 to 1300

NAM: U.S. National Association of Manufacturers

paleoclimatology: the study of past climates, including past climate change

PDO: the Pacific Decadal Oscillation, a natural climate cycle in the Pacific Ocean of much longer duration than El Niño and La Niña

urban heat island: a term referring to the warmth generated by urban surroundings, such as buildings and concrete, that biases measured temperatures upward

UV: ultraviolet, one of three main types of radiation emitted by the sun (ultraviolet, visible and infrared); UV and visible radiation are shortwave, while infrared is longwave

Notes and References

Chapter 1: Global Warming Deceit

1   A recent book that profiles prominent global warming skeptics is titled The Deniers, in order to emphasize alarmists' derogatory use of the term – Lawrence Solomon, *The Deniers: The World Renowned Scientists Who Stood Up Against Global Warming Hysteria, Political Persecution, and Fraud* (Richard Vigilante Books, 2008).

2   Ray Evans, "The Kyoto Protocol's Statistical Fallacies", *Quadrant*, May 2003.

3   Al Gore, *An Inconvenient Truth* (Rodale Press, 2006), pp. 260-263.

4   U.S. Senate Environment and Public Works Committee (Minority) Report, December 20, 2007, found at http://epw.senate.gov/public/index.cfm?FuseAction=Minority. SenateReport.

5   U. S. Senate Minority Report, "More Than 700 International Scientists Dissent Over Man-Made Global Warming Claims: Scientists Continue to Debunk 'Consensus' in 2008 & 2009", updated March 16, 2009, found at http://epw.senate.gov/public/index. cfm?FuseAction=Files.View&FileStore_id=83947f5d-d84a-4a84-ad5d-6e2d71db52d9.

6   Richard S. Lindzen, "Climate of Fear", *The Wall Street Journal*, Opinion Journal, April 12, 2006.

7   Hans von Storch and Dennis Bray, "Climate scientists' views on climate change: A survey", reported on *Nature* blog, Climate Feedback, August 8, 2007, found at http://blogs.nature. com/climatefeedback/2007/08/climate_scientists_views_on_cl_1.html#more.

8   Peter T. Doran and Maggie Kendall Zimmerman, "Examining the scientific consensus on climate change", *Eos Transactions AGU* 90, pp. 22-23 (2009), available at http://tigger.uic. edu/~pdoran/012009_Doran_final.pdf.

9   Global Warming Petition Project, Oregon Institute of Science and Medicine. See http:// www.petitionproject.org/.

10 Naomi Oreskes, "Beyond the ivory tower: The scientific consensus on climate change", *Science* 306, p. 1686 (December 3, 2004) and Erratum, *Science* 307, p. 355 (January 21, 2005). Oreskes' article refers to the keywords "climate change", which would have yielded about 11,000 abstracts in the ISI database for the period from 1993 to 2003. The erratum reports the "global climate change" keywords she actually used, which generated the much smaller number of 928 abstracts.

11 Benny Peiser, "The Dangers of Consensus Science", *National Post*, May 17, 2005, reported at http://www.staff.livjm.ac.uk/spsbpeis/NationalPost.htm. Peiser disputes several of Oreskes' conclusions, including her statement that no abstracts disagreed with the supposed consensus on man-made global warming.

12 Gallup Poll, "Awareness, Opinions About Global Warming Vary Worldwide", April 22, 2009, found at http://www.gallup.com/poll/117772/Awareness-Opinions-Global-Warming-Vary-Worldwide.aspx#1. Alarmists are defined as those polled who think that global warming is a result of human activity. For the world, the alarmist percentage (54%) is the median of the percentages for the 127 countries surveyed, which include the U.S. The survey also revealed that more than a third of the world's population hasn't even heard of global warming.

13 Pew Research Center, January 24, 2007, found at http://pewresearch.org/pubs/282/global-warming-a-divide-on-causes-and-solutions.

14 IPCC, *Climate Change 2007*: The Physical Science Basis, Chapter 1: Historical Overview of Climate Change Science, Section 1.6.

15 Ibid.

16 IPCC, *Climate Change 2007*: The Physical Science Basis, Summary for Policymakers.

17 For an explanation of greenhouse gases, see Chapter 2: The Fuss About CO_2.

18 IPCC, 2007. See http://www.ipcc.ch/pdf/press-ar4/ipcc-flyer-low.pdf.

19 Through photosynthesis, the process by which living cells convert light to chemical energy.

20 Sunspots are small dark spots on the sun's surface which come from magnetic storms, and which can be seen through a filter on a telescope.

21 IPCC, *Climate Change 2007*: The Physical Science Basis, Chapter 1: Historical Overview of Climate Change Science, Section 1.4.3.

22 Cosmic rays are highly energetic charged particles, most of which originate in the shock waves of supernova explosions that occur far away in our galaxy.

Chapter 2: The Fuss About CO_2

23 HadCRUT3 global combined land and sea surface temperature record, from http://www.cru.uea.ac.uk/cru/info/warming/. The Climatic Research Unit (CRU) is part of the University of East Anglia in the UK, and compiles the temperature record jointly with the UK Met Office Hadley Centre.

24 http://climateprediction.net/schools/docs/correlation_causation.ppt#257,5,Data for last 150 years.

25 IPCC, *Climate Change 2007*: The Physical Science Basis, Summary for Policymakers, p. 5. Although there is general agreement that warming has occurred, the exact amount of the temperature increase has been disputed, as discussed in Chapter 3: Science Gone Wrong of this book. As well as the temperature increase itself, the rate of temperature increase since 1980 is in dispute. If the increase is smaller than 0.8° Celsius (1.4° Fahrenheit), the CO_2 hypothesis is less likely to be valid. But in any case, the hypothesis is invalidated by the three red flags for CO_2 that are listed in the lower part of Table 2.1.

26 Today's (2009) CO_2 level in the atmosphere is 387 parts per million. This is 38% higher than its assumed preindustrial level in the 1700s, which was 280 parts per million according to the IPCC (in Reference 25 above, p. 2), or about 35% higher than in 1850. However, the actual magnitude of the CO_2 increase may be considerably lower than these values indicate, as discussed in Chapter 3: Science Gone Wrong.

27 Solar radiation is absorbed and radiated by the Earth and its atmosphere in two different wavelength regions: absorption takes place at short (ultraviolet and visible) wavelengths, while heat is radiated away at long (infrared) wavelengths. Greenhouse gases in the atmosphere allow most of the incoming shortwave radiation to pass through, but absorb a substantial portion of the outgoing longwave radiation.

28 Svante Arrhenius, "Die vermutliche ursache der klimaschwankungen", *Meddelanden från K. Vetenskapsakademiens Nobelinstitut* 1, pp. 1–10 (1906).

29 The atmosphere on Venus, which is very dense, consists of 97% CO_2. This near immersion of the planet in CO_2 contributes to the high and inhospitable surface temperature there of around 470° Celsius (880° Fahrenheit), though some of the greenhouse warming on Venus comes from thick banks of sulfuric acid clouds sitting above the CO_2.

30 Richard S. Lindzen, "Understanding common climate claims", in *Proceedings of the 34th International Seminar on Nuclear War and Planetary Emergencies*, ed. R. Raigaini (World Scientific Publishing Co., 2006), pp. 189-210.

31 Roy W. Spencer, *Climate Confusion: How Global Warming Hysteria Leads to Bad Science, Pandering Politicians and Misguided Policies that Hurt the Poor* (Encounter Books, 2008), p. 65.

Chapter 3: Science Gone Wrong

32 IPCC, *Climate Change 2007*: The Physical Science Basis, Chapter 1: Historical Overview of Climate Change, Section 1.2.

33 The scientific method consists of six basic steps: (1) Observation or data gathering; (2) Formulation of a hypothesis, that is an educated guess, to explain the observations; (3) Initial testing of the hypothesis by experiment; (4) Further, independent testing if the hypothesis is verified by the initial tests; (5) Elevation of the hypothesis, or several hypotheses, to a theory (based on a limited number of observations) or a law (based on many observations); (6) Modification or rejection of any hypothesis if contrary observations are made during experimental testing. In formulating hypotheses, two of the standard reasoning processes used are induction and deduction. Induction reasons from a particular case to a general conclusion, while deduction reasons from general evidence to a particular conclusion – the CO_2 hypothesis is an example of deduction. An important element of the scientific method is the falsifiability criterion, which says that a single contrary observation can invalidate a whole theory or law (though not generally a hypothesis).

34 IPCC, *Climate Change 2007*: The Physical Science Basis, Chapter 1: Historical Overview of Climate Change Science, Section 1.3.3.

35 Bias, a statistician's term, is known as systematic error in scientific parlance.

36 S.A. Changnon, "A rare long record of deep soil temperatures defines temporal temperature changes and an urban heat island", *Climatic Change* 38, pp. 113-128 (1999).

37 Ross R. McKitrick and Patrick J. Michaels, "A test of corrections for extraneous signals in gridded surface temperature data", *Climate Research* 26, pp. 159–173 (2004) and Erratum, *Climate Research* 27, pp. 265– 268 (2004); and "Quantifying the influence of anthropogenic surface processes and inhomogeneities on gridded global climate data", *Journal of Geophysical Research* 112, p. D24S09 (2007), found at http://www.uoguelph.ca/~rmckitri/research/jgr07/M&M.JGRDec07.pdf.

38 A.T.J. De Laat and A.N. Maurellis, "Industrial CO_2 emissions as a proxy for anthropogenic influence on lower tropospheric temperature trends", *Geophysical Research Letters* 31, p. L05204 (2004), and "Evidence for influence of anthropogenic surface processes on lower tropospheric and surface temperature trends", *International Journal of Climatology* 26, pp. 897– 913 (2006).

39 IPCC, *Climate Change 2007*: The Physical Science Basis, Chapter 3: Observations: Surface and Atmospheric Climate Change, Section 3.2.2.2.

40 McKitrick and Michaels (in Reference 37 above) estimate that the land surface warming rate since 1980 falls from the IPCC's value of 0.27° Celsius (0.49° Fahrenheit) per decade to 0.13° Celsius (0.23° Fahrenheit) per decade, when the effect of urban heat islands is

included. This land warming rate can be combined with the IPCC ocean warming rate of 0.13° Celsius (0.23° Fahrenheit) per decade, which is the same value, to give a corrected post-1980 global warming rate also equal to 0.13° Celsius (0.23° Fahrenheit) per decade. This is 24% lower than the IPCC global warming rate of 0.17° Celsius (0.31° Fahrenheit) per decade. The IPCC warming rates (from Reference 39 above, Executive Summary) are weighted averages over the Northern and Southern Hemispheres.

41   If the corrected temperature is lower by one fourth – that is, equal to three fourths of the original, uncorrected temperature – then the original temperature is higher by one fourth of its value, or by one third of the lower, corrected temperature.

42   Howard C. Hayden, *A Primer on CO$_2$ and Climate* (Vales Lake Publishing, 2nd edition, 2008), Figure 11.

43   Satellite global land and sea surface temperature record, reported at http://vortex.nsstc. uah.edu/data/msu/t2lt/uahncdc.lt.

44   See NASA Goddard Institute for Space Studies (GISS) Surface Temperature Analysis, Station Data (for stations with reasonably long, consistently measured time records), reported at http://data.giss.nasa.gov/gistemp/station_data/; and Ross R. McKitrick, at http://www.uoguelph.ca/~rmckitri/research/nvst.html.

45   Craig Loehle and J. Huston McCulloch, "Correction to: A 2000-year global temperature reconstruction based on non-tree ring proxies", *Energy & Environment* 19, pp. 93-100 (2008). The proxies used in this reconstruction included ocean sediments, boreholes, pollen, and cave stalagmites.

46   IPCC, *Climate Change 2007*: The Physical Science Basis, Figure 6.4 and FAQ 2.1, Figure 1.

47   Michael E. Mann, Raymond S. Bradley and Malcolm K. Hughes, "Global-scale temperature patterns and climate forcing over the past six centuries", *Nature* 392, pp. 779-787(1998), and "Northern Hemisphere temperatures during the past millennium: Inferences, uncertainties, and limitations", *Geophysical Research Letters* 26, pp. 759-762 (1999).

48   Stephen McIntyre and Ross R. McKitrick, "Corrections to the Mann et. al. (1998) proxy data base and Northern Hemispheric average temperature series", *Energy & Environment* 14 , pp. 751-771 (2003), discussed further at http://www.uoguelph.ca/~rmckitri/research/ MM-W05-background.pdf.

49   IPCC, *Climate Change 1995*: The Science of Climate Change, Figure 22.

50   IPCC, *Climate Change 2001*: The Scientific Basis, Figure 2.20.

51   See Reference 48 above; principal components is a statistical term.

52 Edward J. Wegman, David W. Scott and Yasmin H. Said, U.S. House Committee on Energy and Commerce Ad Hoc Committee Report on the 'Hockey Stick' Global Climate Reconstruction, July 19, 2006, found at http://www.uoguelph.ca/~rmckitri/research/WegmanReport.pdf.

53 *Surface Temperature Reconstructions for the Last 2,000 Years* (National Academies Press, 2006), Executive Summary, found at http://www.nap.edu/catalog/11676.html.

54 S. Fred Singer and Dennis T. Avery, *Unstoppable Global Warming: Every 1,500 Years* (Rowman & Littlefield, 2nd edition, 2008), pp. 46-92.

55 Michael E. Mann, Zhihua Zhang, Malcolm K. Hughes, Raymond S. Bradley, Sonya K. Miller, Scott Rutherford and Fenbiao Ni, "Proxy-based reconstructions of hemispheric and global surface temperature variations over the past two millennia", *Proceedings of the National Academy of Sciences* 105, pp. 13252–13257 (2008), available at http://www.pnas.org/content/early/2008/09/02/0805721105.full.pdf+html.

56 http://news.bbc.co.uk/2/hi/science/nature/7592575.stm.

57 IPCC, *Climate Change 2007*: The Physical Science Basis, Chapter 6: Palaeoclimate, Box 6.2 and Section 6.6.1.1.

58 David Deming, "Global warming, the politicization of science, and Michael Crichton's *State of Fear*", *Journal of Scientific Exploration* 19, no.4 (2005).

59 Ibid.

60 IPCC, *Climate Change 2007*: The Physical Science Basis, Chapter 2: Changes in Atmospheric Constituents and in Radiative Forcing, Section 2.3.1.

61 Ernst-Georg Beck, "180 years of atmospheric CO_2 gas analysis by chemical methods", *Energy & Environment* 18 , pp. 259-282 (2007). The data that Beck has compiled in Figure 3.5 consists of approximately 70,000 out of more than 90,000 chemical measurements of the CO_2 background level, which are accurate to 3% or better; Beck rejected the other 20,000 because of bias. These measurements come from locations mostly in central and northern Europe, with additional observations from stations in the U.S., India, the Atlantic Ocean and Antarctica.

62 The higher CO_2 readings reflect measurements made in or near large cities, in close proximity to industrial sources of CO_2. Despite the swings, however, the *average* CO_2 level is still meaningful. The average of the chemical data shown in Figure 3.5 exceeds the IPCC's assumed preindustrial baseline for CO_2 by more than 40 parts per million.

63 An astute reader may, however, notice that the chemical data for CO_2 shown in Figure 3.5 show a strong peak around the year 1940, which corresponds almost exactly to a smaller peak in the temperature record shown in Figure 2.1. But this is probably a coincidence, as other peaks in the CO_2 data are not matched in the temperature record. The CO_2 level

in the 1940 peak of about 440 parts per million is well above even the present-day (2009) value of 387 parts per million.

64 G.S. Callendar, "The composition of the atmosphere through the ages," *The Meteorological Magazine* 74, pp. 33-39 (1939).

65 Giles Slocum, "Has the amount of carbon dioxide in the atmosphere changed significantly since the beginning of the twentieth century?", *Monthly Weather Review*, October 1955, pp. 225-231.

66 Ibid.

67 Atmospheric CO_2 levels are determined from ice cores by measuring the composition of air bubbles trapped in the ice, but the measured levels may not be accurate. Because the air isn't captured until enough snow has accumulated to pack the subsurface snow into ice, the age of the trapped air is less than the age of the surrounding ice, but by an amount that varies with local conditions. This introduces uncertainty into the time scale and makes it difficult to precisely match indirect ice-core data with recent, direct measurements of CO_2. And ice-core CO_2 levels may be inaccurate due to changes in the composition of the original air caused by complex physical and chemical processes that occur in the packed ice. Even if that is not the case, the CO_2 level in Antarctica and Greenland, from where most ice cores are extracted, is somewhat lower than the global average owing to absorption of CO_2 by the colder oceans in those regions.

68 See, for example, M. Rundgren and D. Beerling, "A Holocene CO_2 record from the stomatal index of subfossil Salix herbacea L. leaves from northern Sweden", *Holocene* 9, pp. 509-513 (1999).

69 Friederike Wagner, Sjoerd J. P. Bohncke, David L. Dilcher, Wolfram M. Kurschner, Bas van Geel and Henk Visscher, "Century-scale shifts in early Holocene atmospheric CO_2 concentration", *Science* 284, pp. 1971-1973 (1999); and Thomas B. van Hoof, Friederike Wagner-Cremer, Wolfram M. Kurschner and Henk Visscher, "A role for atmospheric CO_2 in preindustrial climate forcing", *Proceedings of the National Academy of Sciences* 105, pp. 15815–15818 (2008).

70 Current CO_2 background levels in the atmosphere are monitored at about 50 sites around the world by the U.S. Department of Energy (DOE)'s Carbon Dioxide Information Analysis Center (CDIAC), reported at http://cdiac.ornl.gov/trends/co2/contents.htm. Most man-made CO_2 is generated in the Northern Hemisphere, but quickly spreads across the globe.

71 The IPCC (in Reference 39 above, Executive Summary) claims that the warming rate from urban heat islands is only 0.006° Celsius (0.01° Fahrenheit) per decade, compared with the overall land surface warming rate since 1980 of 0.27° Celsius (0.49° Fahrenheit) per decade – or a negligible 2.2%. This contrasts with the estimate made by McKitrick and Michaels

(in Reference 37 above) that the urban heat island effect contributes 0.14° Celsius (0.25° Fahrenheit) of the 0.27° Celsius (0.49° Fahrenheit) per decade warming rate – more than 50% of the measured temperature increase on land.

72 Ross R. McKitrick and Patrick J. Michaels, Background discussion on "Quantifying the influence of anthropogenic surface processes and inhomogeneities on gridded global climate data", found at http://www.uoguelph.ca/~rmckitri/research/jgr07/M&M.JGR07-background.pdf.

73 Ibid.

74 IPCC, *Climate Change 2007*: The Physical Science Basis, Expert and Government Review Comments on the Second-Order Draft, Chapter 3, No. 3-285, found at http://pds.lib.harvard.edu/pds/view/7786376?n=38&imagesize=1200&jp2Res=.25.

75 Ross R. McKitrick, "Atmospheric oscillations do not explain the temperature-industrialization correlation", Social Science Research Network, available at: http://papers.ssrn.com/sol3/papers.cfm?abstract_id=1166424.

76 Douglas J. Keenan, "The fraud allegation against some climatic research of Wei-Chyung Wang", *Energy & Environment* 18 , pp. 985-995 (2007), summarized and updated at http://www.informath.org/apprise/a5620.htm.

77 Wei-Chyung Wang, Zhaomei Zeng and Thomas R. Karl, "Urban heat islands in China", *Geophysical Research Letters* 17, pp. 2377–2380 (1990).

78 P.D. Jones, P.Y. Groisman, M. Coughlan, N. Plummer, W.-C. Wang and T.R. Karl, "Assessment of urbanization effects in time series of surface air temperature over land", *Nature* 347, pp. 169–172 (1990).

79 Tao Shiyan, Fu Congbin, Zeng Zhaomei, and Zhang Qingyun, "Two long-term instrumental climatic data bases of the People's Republic of China", Oak Ridge National Laboratory report ORNL/CDIAC-47, NDP-039 (1991), available at http://cdiac.ornl.gov/ftp/ndp039/ndp039.pdf.

80 Douglas J. Keenan (September 2007), at http://www.informath.org/apprise/a5620/b17.htm.

81 IPCC, *Climate Change 2007*: The Physical Science Basis, Chapter 3: Observations: Surface and Atmospheric Climate Change, Section 3.2.2.2. Phil Jones, lead author of Reference 78 above, was also one of the two lead authors of this IPCC report chapter.

82 P.D. Jones, D. H. Lister and Q. Li, "Urbanization effects in large-scale temperature records, with an emphasis on China", *Journal of Geophysical Research* 113, p. D16122 (2008). The study concludes that urban-related warming in China was about 0.1° Celsius (0.2° Fahrenheit) per decade during the period 1951–2004. Over 5.4 decades, this amounts to a contribution from urbanization of 0.54° Celsius (0.97° Fahrenheit). This contribution is

two thirds of the true climatic warming of 0.81° Celsius (1.5° Fahrenheit) over the same period.

83   Richard S. Lindzen, "Climate Science: Is it currently designed to answer questions?", found at http://arxiv.org/ftp/arxiv/papers/0809/0809.3762.pdf.

84   IPCC, *Appendix A to the Principles Governing IPCC Work* (amended November 2003), Section 4.2, found at http://www.ipcc.ch/pdf/ipcc-principles/ipcc-principles-appendix-a. pdf.

85   Vincent R. Gray, letter to David Henderson (2007), found at http://nzclimatescience.net/ index.php?option=com_content&task=view&id=155&Itemid=1.

86   Ibid.

87   S. Fred Singer, Letter to IPCC (Working Group I) Scientists, found at http://www.sepp.org/ Archive/controv/ipcccont/ipccflap.htm.

88   IPCC, *Climate Change 1995*: The Science of Climate Change, Chapter 8, p. 412 and p. 439.

89   Frederick Seitz, "A Major Deception on Global Warming", *The Wall Street Journal*, June 12, 1996.

90   Editorial, "Climate debate must not overheat", *Nature* 381, p. 539 (1996).

91   Paul N. Edwards and Stephen H. Schneider, "The 1995 IPCC Report: Broad consensus or 'scientific cleansing'?", *Ecofable/Ecoscience* 1, pp. 3-9 (1997), found at http://www.si.umich. edu/~pne/PDF/ecofables.pdf.

92   IPCC, *Climate Change 2007*: The Physical Science Basis, Contribution of Working Group I to the Fourth Assessment Report of the IPCC, Preface.

93   John McLean, "Why the IPCC should be disbanded", November 9, 2007, found at http:// scienceandpublicpolicy.org/originals/whytheipccshouldbedisbanded.html.

94   IPCC, *Climate Change 2007*: The Physical Science Basis, Expert and Government Review Comments on the Second-Order Draft, Chapter 9, found at http://pds.lib.harvard.edu/ pds/view/7787808?n=1&imagesize=1200&jp2Res=.25.

95   IPCC, *Appendix A to the Principles Governing IPCC Work* (amended November 2003), Section 4.2.5, found at http://www.ipcc.ch/pdf/ipcc-principles/ipcc-principles-appendix-a. pdf.

Chapter 4: Computer Snake Oil?

96   Charles Blilie, *The Promise And Limits Of Computer Modeling* (World Scientific, 2007), Chapter 1.

97   IPCC, *Climate Change 2007*: The Physical Science Basis, Chapter 1: Historical Overview of Climate Change, Section 1.2.

98 CCSP, 2008: Climate Models: An Assessment of Strengths and Limitations, Report by the U.S. Climate Change Science Program and the Subcommittee on Global Change Research, Department of Energy (DOE), Office of Biological and Environmental Research, Section 1.2.

99 Examples of processes or quantities represented by adjustable parameters in climate models include flows of the atmosphere and oceans; atmospheric turbulence near the earth's surface (in the so-called boundary layer); the relative humidity threshold for cloud formation (separate sets of parameters for low-level and high-level clouds); the efficiency of precipitation evaporation; aerosol microphysics; ocean salinity; sea-ice dynamics; and many more.

100 Quoted by Freeman Dyson in "A meeting with Enrico Fermi", *Nature* 427, p. 297 (2004).

101 James M. Murphy, David M. H. Sexton, David N. Barnett, Gareth S. Jones, Mark J. Webb, Matthew Collins and David A. Stainforth, "Quantification of modelling uncertainties in a large ensemble of climate change simulations", *Nature* 430, pp. 768-772 (2004).

102 Global energy balance refers to the balance between the amount of absorbed sunlight at the top of the atmosphere and the amount of energy that the Earth radiates away to space, both of which are estimated to be about 235 watts per square meter when averaged over a year. It is this balance that is believed to keep our planet at a constant average temperature over the short term. Computer climate models are tuned by adjusting parameters, typically for clouds, to maintain this energy balance. However, because 235 watts per square meter is only an inexact theoretical calculation, the tuning procedure is also subject to uncertainty. This uncertainty can affect the model's ability to correctly predict climate sensitivity, as discussed by Bender [F. A.-M. Bender, *Environmental Research Letters* 3, pp. 1-6 (2008), found at http://www.iop.org/EJ/article/1748-9326/3/1/014001/erl8_1_014001.html].

103 CCSP, 2008: Climate Models: An Assessment of Strengths and Limitations, Report by the U.S. Climate Change Science Program and the Subcommittee on Global Change Research, Department of Energy (DOE), Office of Biological and Environmental Research, Section 2.5.2, discussing the Community Climate System Model – one of the major classes of climate model used by IPCC modelers.

104 As described in IPCC, *Climate Change 2007*: The Physical Science Basis, Chapter 8: Climate Models and Their Evaluation.

105 Kesten C. Green and J. Scott Armstrong, "Global warming: Forecasts by scientists versus scientific forecasts", *Energy & Environment* 18 , pp. 997-1021 (2007), found at http://www.forecastingprinciples.com/Public_Policy/WarmAudit31.pdf.

106 Ibid.

107 The main uses of adjustable cloud parameters are to approximate cloud formation and dissipation for cirrus (high-level) and stratus clouds; cumulus (low-level) cloud convection, in both fair weather and thunderstorms; turbulence; and conversions between cloud water, rainwater, ice crystals, and snow.

108 "Cloud-resolving models", which have grid boxes less than a few kilometers square, are already available, but take too long to run even on supercomputers today. These models can simulate features such as deep updrafts and downdrafts, cirrus anvil clouds, and aerosol-cloud interactions better than present-day climate models, though still require parameterization of many subgrid-scale processes.

109 James Hansen et al, "Climate simulations for 1880–2003 with GISS model E", *Climate Dynamics* 29, pp. 661–696 (2007). W/m^2 (watts per square meter) is a unit of radiative energy; hPa (hectopascals) is a unit of atmospheric pressure, standard atmospheric pressure at sea level being 1013 hPa.

110 United Nations Framework Convention on Climate Change (UNFCCC), Convention Text (1992), Article 2, found at http://unfccc.int/resource/docs/convkp/conveng.pdf.

111 IPCC, *Climate Change 2007*: The Physical Science Basis, Chapter 8: Climate Models and Their Evaluation, FAQ 8.1.

112 Ibid, Section 8.6.3.2.

113 IPCC, *Climate Change 2007*: The Physical Science Basis, Chapter 10: Global Climate Projections, Section 10.2.1.3 and Figure 10.11(a).

114 IPCC, *Climate Change 2007*: The Physical Science Basis, Chapter 9: Understanding and Attributing Climate Change, Table 9.4.

115 Computer models are unable to predict the timing (except for El Niño) and climatic effects of either the El Niño-Southern Oscillation, which is the warm phase of an ocean-atmosphere cycle occurring at irregular intervals of 2-8 years, or the Pacific Decadal Oscillation, or the intraseasonal Madden-Julian Oscillation.

116 This is one of several positive feedback mechanisms that could enhance global warming in polar regions, if the CO_2 hypothesis were valid. The feedback mechanism involves warmth from CO_2 melting snow and ice to expose darker surfaces with lower reflectivity (albedo). The less reflective surfaces absorb more solar heat, enhancing the warming from CO_2 alone.

117 Igor V. Polyakov et al, "Variability and trends of air temperature and pressure in the maritime Arctic", *Journal of Climate* 16, pp. 2067-2077 (2003).

118 Ibid.

119 Peter T. Doran et al, "Antarctic climate cooling and terrestrial ecosystem response", *Nature* 415, pp. 517-520 (2002).

120 Eric J. Steig, David P. Schneider, Scott D. Rutherford, Michael E. Mann, Josefino C. Comiso and Drew T. Shindell, "Warming of the Antarctic ice-sheet surface since the 1957 International Geophysical Year", *Nature* 457, pp. 459-462 (2009).

121 Hu McCulloch, "Steig 2009's Non-Correction for Serial Correlation", February 26, 2009, found at http://www.climateaudit.org/?p=5341. The correction involves what statisticians call serial correlation. Because of the limited number of Antarctic weather stations supplying actual temperature data, the much larger number of reconstructed temperatures across the continent are correlated. This correlation increases the statistical errors beyond those reported by Steig et al (in Reference 120 above). When the correlation is taken into account, the statistical error in the reconstructed temperature for East Antarctica becomes greater than 100% – meaning that the temperature change over the past 50 years could be as small as zero.

122 John Turner et al, "Non-annular atmospheric circulation change induced by stratospheric ozone depletion and its role in the recent increase of Antarctic sea ice extent", *Geophysical Research Letters* 36, p. L08502 (2009). The study of Antarctic sea ice was conducted by the British Antarctic Survey and NASA.

123 IPCC, *Climate Change 2007*: The Physical Science Basis, Chapter 7: Couplings Between Changes in the Climate System and Biogeochemistry, Section 7.5.2.4. Aerosol particles in the atmosphere tend to remain in the Northern Hemisphere where they are produced, unlike greenhouse gases such as CO_2, which quickly become mixed between the two hemispheres.

124 IPCC, *Climate Change 2007*: The Physical Science Basis, Chapter 3: Observations: Surface and Atmospheric Climate Change, Section 3.2.2.4.

125 Specifically, the troposphere – the lowest layer of the atmosphere.

126 David H. Douglass, John R. Christy, Benjamin D. Pearson and S. Fred Singer, "A comparison of tropical temperature trends with model predictions", *International Journal of Climatology* 28, pp. 1693–1701 (2008).

127 John M. Lyman, Josh K. Willis and Gregory C. Johnson, "Recent cooling of the upper ocean", *Geophysical Research Letters* 33, p. L18604 (2006); and Josh K. Willis, John M. Lyman, Gregory C. Johnson and John Gilson, "Correction to 'Recent cooling of the upper ocean'", *Geophysical Research Letters* 34, p. L16601 (2007).

128 IPCC, *Climate Change 2007*: The Physical Science Basis, Chapter 8: Climate Models and Their Evaluation, Section 8.3.2.

Chapter 5: CO$_2$ Sense and Sensitivity

129 Typically, between barometric pressure and wind speed – with lower pressure increasing wind flow, which drops the pressure more and causes even higher winds, and so on.

130 For water vapor, which is the dominant greenhouse gas in the atmosphere, the idea is that a small increase in atmospheric CO$_2$ warms the Earth slightly via the greenhouse effect. This warming produces more water vapor by evaporation from oceans and lakes, and the extra water vapor then amplifies the warming even more. For clouds, the initial CO$_2$ warming produces more high-level clouds that cause further warming (as opposed to low-level clouds that cause cooling). Melting of terrestrial snow and sea ice by CO$_2$ warming exposes darker surfaces such as soil, rock and seawater, which have lower albedo. The less reflective surfaces absorb more of the sun's radiation and thus push temperatures higher. In all three cases, the tiny effect from CO$_2$ acting alone is magnified, according to climate modelers and the IPCC.

131 IPCC, *Climate Change 2007*: The Physical Science Basis, Chapter 8: Climate Models and Their Evaluation, Section 8.6.

132 IPCC, *Climate Change 2007*: The Physical Science Basis, Chapter 7: Couplings Between Changes in the Climate System and Biogeochemistry, Section 7.3.5.

133 Ibid, Section 7.6.

134 The water vapor feedback, which depends on CO$_2$ through the greenhouse effect, is the strongest feedback in IPCC climate models. However, its strength depends on the assumption that global warming does not change the relative humidity in the lower atmosphere (troposphere). Warming increases the water vapor concentration (specific humidity) in the atmosphere, but also increases the saturation level for water vapor, which defines relative humidity. But if the relative humidity distribution in the troposphere does not remain constant as warming occurs, the water vapor feedback may be weaker than estimated.

135 Panel on Climate Change Feedbacks, National Research Council, *Understanding Climate Change Feedbacks* (National Academies Press, 2003), p. 24. The technical term for the rate of decrease of temperature with altitude in the troposphere is the lapse rate. The greenhouse effect for CO$_2$ is enhanced or diminished by a higher or lower lapse rate, respectively, since water vapor in the colder, upper troposphere radiates less heat away from the Earth than water vapor closer to the warmer surface. In the tropics, where temperatures in the troposphere vary most strongly, the observed lapse rate is close to the moist adiabatic lapse rate and has been found to decrease with increasing surface temperature – so that the lapse rate feedback is negative, at least in the tropics. In most IPCC climate models, the lapse rate feedback is also negative, although it is positive in some models. Because temperature and

water vapor changes are so tightly coupled, the positive water vapor feedback is generally combined with the lapse rate feedback.

136 Primarily, from NASA's Tropical Rainfall Measuring Mission (TRMM) and Terra satellites.

137 Roy W. Spencer, William D. Braswell, John R. Christy and Justin Hnilo, "Cloud and radiation budget changes associated with tropical intraseasonal oscillations", *Geophysical Research Letters* 34, p. L15707 (2007). The short–term, intraseasonal climate cycle studied was the Madden-Julian Oscillation. The increase in atmospheric temperature from the cool to the warm phase of this oscillation is, over a few weeks, about as large as the average global warming observed since 1850.

138 Phil Gentry, "Cirrus disappearance: Warming might thin heat-trapping clouds", News Center, University of Alabama in Huntsville, August 9, 2007, found at http://www.uah. edu/News/newsread.php?newsID=875. The heat-trapping ability of high-altitude ice clouds exceeds their solar shading effect, which is the dominant feature of low-altitude clouds.

139 Richard S. Lindzen, Ming-Dah Chou and Arthur Y. Hou, "Does the Earth have an adaptive infrared iris?", *Bulletin of the American Meteorological Society* 82, pp. 417-432 (2001).

140 NASA's Aqua satellite.

141 Roy W. Spencer, "Satellite and climate model evidence against substantial manmade climate change", available at http://www.drroyspencer.com/research-articles/satellite-and-climate-model-evidence/.

142 Roy W. Spencer, private communication (2009). The negative feedback in this second study was not observed in outgoing longwave (infrared) radiation from the Earth, as it was in the first study, but rather in reflected shortwave (ultraviolet) solar radiation. Spencer says that the satellite observations showed an excellent match with the IPCC climate models in the longwave region, suggesting that the total longwave feedback – which consists of the water vapor, lapse rate and high cloud feedbacks – is weakly positive. But this weak positive feedback is dominated by the strong negative feedback from low clouds in the shortwave region. The possibility of zero or even negative longwave feedback, also based on satellite observations, had been suggested earlier by Piers Forster and Jonathan Gregory [Piers M. de F. Forster and Jonathan M. Gregory, "The climate sensitivity and its components diagnosed from Earth radiation budget data", *Journal of Climate* 19, pp. 39-52 (2006)].

143 IPCC, *Climate Change 2007*: The Physical Science Basis, Chapter 8: Climate Models and Their Evaluation, Section 8.6.3.2.2. In roughly half of the current IPCC climate models, the feedback from low-level tropical clouds in response to global warming is negative, which is the same as found from satellite data in the second University of Alabama study (see Reference 141 above). But the other half of the IPCC models predict positive feedback

from low-level clouds, and all models predict net positive feedback from low-level and high-level clouds combined.

144 In Roy Spencer's opinion, current IPCC climate models should be adjusted to mimic the short-term (five-year) satellite data , showing natural cooling associated with negative cloud feedbacks, before making any predictions about long-term global warming.

145 Numerically, the largest feedback parameter in IPCC climate models is the combined water vapor-lapse rate feedback, which is estimated to be 0.96 W/m^2 per °C, compared to the cloud feedback of 0.69 W/m^2 per °C, and the snow and ice feedback of 0.26 W/m^2 per °C (Reference 143 above, Section 8.6.2.3). All these are positive feedbacks.

146 IPCC, *Climate Change 2007*: The Physical Science Basis, Chapter 9: Understanding and Attributing Climate Change, Sections 9.2 and 9.6.1.

147 IPCC, *Climate Change 2007*: The Physical Science Basis, Summary for Policymakers, p. 12.

148 Ibid.

149 Details of these climate sensitivity calculations, and references, are supplied in Appendix A: Climate Feedbacks and Sensitivity.

150 The increase in global surface temperatures since 1850 is 0.6° Celsius (1.1° Fahrenheit) when the urban heat island effect is taken into account, as discussed in Chapter 3: Science Gone Wrong. According to the IPCC (*Climate Change 2007*: The Physical Science Basis, Summary for Policymakers, p. 5), the increase since 1850 is 0.8° Celsius (1.4° Fahrenheit).

151 The preindustrial CO_2 level was 280 parts per million according to the IPCC (*Climate Change 2007*: The Physical Science Basis, Summary for Policymakers, p. 2). However, there is evidence that the preindustrial baseline was higher, at around 320 parts per million, as discussed in Chapter 3: Science Gone Wrong. With this higher baseline, the predicted temperature increase today due to CO_2 would be only 0.32° Celsius with no CO_2 feedback, instead of 0.54° Celsius (Table 5.2), and even less with negative feedback.

152 IPCC, *Climate Change 2007*: The Physical Science Basis, Chapter 9: Understanding and Attributing Climate Change, Section 9.6.4.

153 Dan Pangburn, "Historical Data On Global Warming provided by U.S. Government Agencies", March 15, 2008, found at http://www.middlebury.net/op-ed/pangburn.html.

154 Specifically, ice cores from the Russian Vostok station in East Antarctica.

155 Hubertus Fischer, Martin Wahlen, Jesse Smith, Derek Mastroianni and Bruce Deck, "Ice core records of atmospheric CO_2 around the last three glacial terminations", *Science* 283, pp. 1712-1714 (1999).

156 Nicolas Caillon, Jeffrey P. Severinghaus, Jean Jouzel, Jean-Marc Barnola, Jiancheng Kang and Volodya Y. Lipenkov, "Timing of atmospheric CO_2 and Antarctic temperature changes across termination III", *Science* 299, pp. 1728-1731 (2003).

157 Lowell Stott, Axel Timmermann and Robert Thunell, "Southern Hemisphere and deep-sea warming led deglacial atmospheric CO_2 rise and tropical warming", *Science* 318, pp. 435-438 (2007). The amount of CO_2 released from or absorbed by the oceans as the temperature rises or falls, respectively, is thought to depend on temperature-induced changes both in CO_2 solubility and in ocean mixing processes for CO_2.

158 The orbital changes are known as Milankovitch cycles. See, for example, "Milankovitch Cycles and Glaciation" at http://www.homepage.montana.edu/~geol445/hyperglac/time1/milankov.htm.

159 IPCC, *Climate Change 2007*: The Physical Science Basis, Chapter 6: Palaeoclimate, Box 6.2 and Section 6.4.1.2.

160 The CO_2 feedback mechanisms that amplify cooling are the opposite of those that magnify warming (described in Reference 130 above). The idea is that a small decrease in atmospheric CO_2, caused by the temperature drop at the onset of an ice age, sets off further cooling through a reverse greenhouse effect. Another major, but slower feedback is thought to have come from growth of ice sheets, which are more reflective and absorb less sunlight than the water that froze to produce them. The lower absorption enhances the cooling and causes even more ice to form. At the end of an ice age, the ice sheet feedback enhances warming as the ice melts.

161 The continued rise of the CO_2 level beyond the end of the most recent ice age, and the CO_2 time lag, are barely visible in Figure 5.1 because of the compressed time scale. But they can be clearly seen at the termination of the previous ice age, on the right of the graph.

162 The same problem arises at the beginning of an ice age. Orbital changes can explain the initial temperature dip before atmospheric CO_2 took a downward turn 600-800 years later, as it was sucked into the oceans. But the CO_2 hypothesis has no explanation for what stopped the temperature from continuing to tumble at the end of its decline, while CO_2 kept falling for another 600-800 years.

163 IPCC, *Climate Change 2007*: The Physical Science Basis, Chapter 6: Palaeoclimate, Sections 6.4.1.1 and 6.4.1.2.

164 Barbara Stenni et al, "An oceanic cold reversal during the last deglaciation", *Science* 293, pp. 2074-2077 (2001) – see Fig. 3A. The transition from the last ice age to the present interglacial period took 6,000 to 7,000 years, during which the increase in average global temperature, which is approximately half the change measured at the poles, was about 4° Celsius (7° Fahrenheit). This was six to eight times slower than recent global warming,

the actual ratio depending on the exact magnitude of the temperature increase since 1850 (see Reference 150 above). However, it is well-known that the post ice-age warming was interrupted by an event called the Younger Dryas that resulted in temporary cooling. The warming both before and after the Dryas event was faster, the warming following the event being only two to three times slower than modern global warming.

Chapter 6: Doing What Comes Naturally

165 Known technically as the sun's irradiance, which is measured as the total solar energy incident on the top of the Earth's atmosphere.

166 Shahinaz M. Yousef, "The solar Wolf-Gleissberg cycle and its influence on the Earth", September, 2000, found at http://virtualacademia.com/pdf/cli267_293.pdf.

167 The two principal cosmogenic isotopes are [10]Be, found in ice cores, and ^{14}C, generally measured in tree rings. Because greater solar activity reduces the number of cosmic rays available for atmospheric production of cosmogenic isotopes, which are then transported to the Earth's surface, there is an inverse relation between the sun's output and terrestrial isotope levels. Higher solar activity results in lower isotope levels, and vice versa.

168 IPCC, *Climate Change 2007*: The Physical Science Basis, Chapter 2: Changes in Atmospheric Constituents and in Radiative Forcing, Section 2.7.1.2.1.

169 Y.-M. Wang, J.L. Lean and N.R. Sheeley, Jr., "Modeling the sun's magnetic field and irradiance since 1713", *The Astrophysical Journal* 625, pp. 522–538 (2005).

170 Ibid.

171 IPCC, *Climate Change 2007*: The Physical Science Basis, Chapter 6: Palaeoclimate, Section 6.6.3.4. The IPCC's estimates of the increase in solar irradiance since the time of the Maunder Minimum are: its selected low estimate of 0.08%, based on sunspot data and a computer model of the sun (see Reference 169 above); and a high estimate of 0.25%, derived from a [10]Be ice-core record (see Reference 172 below). The 0.08% gain based on sunspot numbers consists of 0.04% from an increase in the average irradiance, plus half the current sunspot cycle amplitude of 0.08% of the irradiance. The sunspot cycle amplitude during the Maunder Minimum was close to zero.

172 Edouard Bard, Grant Raisbeck, Françoise Yiou and Jean Jouzel, "Solar irradiance during the last 1200 years based on cosmogenic nuclides", *Tellus* 52B, pp. 985-992 (2000). In this study, most of the higher estimates of the increase in solar irradiance since the Maunder Minimum come from now discredited stellar data, but there is one higher estimate by George Reid that is based on sunspot numbers combined with a simple energy balance climate model [George C. Reid, "Solar forcing of global climate change since the mid-17th century", *Climatic Change* 37, pp. 391-405 (1997)]. Reid calculates that solar irradiance has increased by 0.65% from the Maunder Minimum to the present.

173 IPCC, *Climate Change 2007*: The Physical Science Basis, Chapter 9: Understanding and Attributing Climate Change, Figure 9.4.

174 IPCC, *Climate Change 2001*: The Scientific Basis, Chapter 6: Radiative Forcing of Climate Change, Section 6.11.1.2.

175 The closest the 2007 IPCC report (in Reference 173 above, Figure 9.5b) comes to calculating the solar contribution is by graphing the temperature predicted by its climate models over the period from 1900 to the present, with natural forcings alone. The only natural forcings considered are volcanic and solar, but volcanic forcing (which is negative, due to the aerosols released) is currently negligible since there has not been an explosive volcanic eruption since Mt. Pinatubo in 1991. Therefore, the IPCC's predicted temperature increase from natural forcings can be assumed to come just from the sun.

176 IPCC, *Climate Change 2007*: The Physical Science Basis, Chapter 2: Changes in Atmospheric Constituents and in Radiative Forcing, Section 2.7.1.1.2.

177 Claus Fröhlich, "Solar constant: Construction of a Composite Total Solar Irradiance (TSI) Time Series from 1978 to present", Physikalisch-Meteorologisches Observatorium Davos/ World Radiation Center (PMOD/WRC), at http://www.pmodwrc.ch/pmod.php?topic=tsi/ composite/SolarConstant.

178 Richard C. Willson, "Total Solar Irradiance (TSI) Monitoring and Requirements for Sustaining the TSI Database", Active Cavity Radiometer Irradiance Monitor (ACRIM), at http://acrim.com/TSI%20Monitoring.htm.

179 Richard C. Willson, letter to Nicola Scafetta (September 16, 2008), in slide 16 of Scafetta's presentation to the U.S. Environmental Protection Agency (EPA), February 26, 2009, available at http://yosemite.epa.gov/ee/epa/wpi.nsf/09133da7fb9a95db85256698006641d1 /7a5516152467a30b85257562006c89a6!OpenDocument.

180 Nicola Scafetta and Richard C. Willson, "ACRIM-gap and TSI trend issue resolved using a surface magnetic flux TSI proxy model", *Geophysical Research Letters* 36, p. L05701 (2009).

181 Joanna D. Haigh, "The effects of solar variability on the Earth's climate", *Philosphical Transactions of the Royal Society of London* A361, pp. 95-111 (2003).

182 Nicola Scafetta and Bruce J. West, "Phenomenological reconstructions of the solar signature in the Northern Hemisphere surface temperature records since 1600", *Journal of Geophysical Research* 112, p. D24S03 (2007); and "Is climate sensitive to solar variability?", *Physics Today* 61, Issue 3, pp. 50-51 (2008). The basis for the authors' phenomenological model is that short-term variations in the sun's activity over the 11-year solar cycle induce similar fluctuations in the Earth's average temperature – fluctuations that are normally regarded as noise and averaged out in computer climate models. Daily variations in solar activity are associated with sunspots and their bright counterparts, solar flares; monthly

variations occur because of the sun's rotation. Scafetta and West maintain that the short-term fluctuations in the Earth's temperature are not noise, but result in long-term increases and decreases in global temperature over tens or even hundreds of years. In this approach, a significant fraction of present global warming stems from the long-term gain in solar activity since the preindustrial era. Scafetta and West add that most sun-climate coupling mechanisms are not included in computational models, which therefore underestimate solar effects on climate.

183 W. Dansgaard et al, "Evidence for general instability of past climate from a 250-kyr ice-core record", *Nature* 364, pp. 218–220 (1993). Ancient temperatures are deduced from ice cores by measuring the isotopic ratio of ^{18}O to ^{16}O, the ratio being temperature dependent.

184 Gerard Bond et al, "A pervasive millennial-scale cycle in North Atlantic Holocene and glacial climates", *Science* 278, pp. 1257-1266 (1997). The deep-sea sediment cores that were studied contain glacial debris rafted into the oceans by icebergs, and then dropped onto the sea floor as the icebergs melted. Radiocarbon dating (which uses the cosmogenic isotope ^{14}C) of the cores was used to identify the cycle length. The volume of glacial debris is largest, and it is carried farthest out to sea, at the coldest point of the 1,500-year cycle.

185 Ibid.

186 S. Fred Singer and Dennis T. Avery, *Unstoppable Global Warming: Every 1,500 Years* (Rowman & Littlefield, 2nd edition, 2008), pp. 15-99.

187 Holger Braun et al, "Possible solar origin of the 1,470-year glacial climate cycle demonstrated in a coupled model", *Nature* 438, pp. 208-211 (2005).

188 Ibid. To fit an integer multiple of cycles into 1,470 years, the shorter solar cycle would need to be about 86.5 (=1,470/17) years long.

189 Ibid.

190 Gerard Bond et al, "Persistent solar influence on North Atlantic climate during the Holocene", *Science* 294, pp. 2130-2136 (2001). The research team used the cosmogenic isotopes ^{10}Be in Greenland ice cores and ^{14}C in tree rings as proxy indicators of solar activity.

191 M. Debret et al, "The origin of the 1500-year climate cycles in Holocene North-Atlantic records", *Climate of the Past Discussions* 3, pp. 679-692 (2007), found at http://www.clim-past-discuss.net/3/679/2007/cpd-3-679-2007-print.pdf.

192 Raimund Muscheler, Fortunat Joos, Jürg Beer, Simon A. Müller, Maura Vonmoos and Ian Snowball, "Solar activity during the last 1000 yr inferred from radionuclide records", *Quaternary Science Reviews* 26, pp. 82–97 (2007).

193 IPCC, *Climate Change 2007: The Physical Science Basis*, Chapter 6: Palaeoclimate, Section 6.5.1.6.

194 It's actually a lot more complicated than this. Some of the incoming energy from the sun is reflected without being absorbed at all, and much of the absorbed energy is later radiated away – some of it being reabsorbed in the atmosphere, and the rest released into space. Furthermore, most of the absorbed incoming solar radiation is shortwave (ultraviolet and visible), whereas outgoing radiation from the Earth is mostly longwave (infrared).

195 For a discussion of feedback, see Chapter 5: CO_2 Sense and Sensitivity. In addition to the cosmic ray and ozone heating mechanisms discussed here in Chapter 6: Doing What Comes Naturally, any positive feedbacks that magnify the warming effect of CO_2 will also magnify solar heating effects.

196 Known as the "solar wind".

197 Henrik Svensmark, "Cosmoclimatology: A new theory emerges", *Astronomy & Geophysics* 48, pp. 1.18-1.24 (2007); Henrik Svensmark and Nigel Calder, *The Chilling Stars: A Cosmic View of Climate Change* (Totem Books, 2nd edition, 2008). The proposed mechanism for cloud formation involves cosmic ray ionization of air molecules, resulting in the production of aerosol particles that consist of ultrasmall clusters of sulfuric acid and water molecules. Svensmark says that these clusters are the building blocks of cloud condensation nuclei. He presents data showing a strong correlation between low cloud cover and cosmic rays reaching the Earth's atmosphere, over the last 25 years.

198 Ibid.

199 Ibid.

200 Nir J. Shaviv and Ján Veizer, "Celestial driver of Phanerozoic climate?", *Geological Society of America Today* 13, pp. 4-10 (2003).

201 The small-scale experiment, conducted at the Danish National Space Center in Copenhagen, simulated conditions only at sea level, where the cosmic ray flux is mostly muons. Muons are secondary cosmic rays formed by the interaction of primary cosmic ray particles (such as protons and helium nuclei) with air molecules higher in the atmosphere. The full-scale experiment, to be carried out at CERN in Geneva, will use a high-energy particle beam to simulate primary cosmic rays, and the cloud chamber will be able to simulate conditions at all levels of the atmosphere. The larger cloud chamber should also make it possible to produce actual cloud condensation nuclei, rather than just their building blocks as seen in the small-scale chamber.

202 IPCC, *Climate Change 2007*: The Physical Science Basis, Chapter 2: Changes in Atmospheric Constituents and in Radiative Forcing, Section 2.7.1.3.

203 Ibid.

204 Ultraviolet (UV) radiation from the sun breaks up oxygen (O_2) molecules in the upper atmosphere (stratosphere) into individual atoms of oxygen, which then combine with

unsplit O_2 to form ozone (O_3). The ozone layer absorbs almost all the sun's UVB radiation, which causes sunburn and, at high doses, skin cancer. In addition to absorbing UV from the sun, ozone is a greenhouse gas like CO_2 and traps a small portion of the infrared heat radiated away from the Earth.

205 Drew Shindell, David Rind, Nambeth Balachandran, Judith Lean and Patrick Lonergan, "Solar cycle variability, ozone, and climate", *Science* 284, pp. 305-308 (1999).

206 See Reference 181 above.

207 Ibid.

208 Joanna D. Haigh, Henrik Lundstedt, Peter Wintoft, Henrik Svensmark, Freddy Christiansen and Nigel Marsh, "Influence of Solar Activity Cycles on Earth's Climate", presentation at Second European Space Weather Week, November 15, 2005, found at http://www.esa-spaceweather.net/spweather/workshops/eswwII/proc/Session2/ESTECsww_20051.pdf. Apart from cosmic rays and solar UV, the main alternative mechanism for amplifying the sun's activity involves the solar wind.

209 Paul E. Geissler, "Three decades of Martian surface changes", *Journal of Geophysical Research* 110, p. E02001 (2006); Lori K. Fenton, Paul E. Geissler and Robert M. Haberle, "Global warming and climate forcing by recent albedo changes on Mars", *Nature* 446, pp. 646-649 (2007). The warming was predicted by a computer climate model for Mars, in which the darker surface absorbed more sunlight than the lighter, dusty surface 22 years earlier. The ice caps on Mars are solid CO_2, not water ice as on the Earth.

210 Mark A. Szwast, Mark I. Richardson and Ashwin R. Vasavada, "Surface dust redistribution on Mars as observed by the Mars Global Surveyor and Viking orbiters", *Journal of Geophysical Research* 111, p. E11008 (2006).

211 L.A. Sromovsky, P.M. Fry, S.S. Limaye and K.H. Baines, "The nature of Neptune's increasing brightness: Evidence for a seasonal response", *Icarus* 163, pp. 256–261 (2003).

212 H.B. Hammel and G.W. Lockwood, "Suggestive correlations between the brightness of Neptune, solar variability, and Earth's temperature", *Geophysical Research Letters* 34, p. L08203 (2007).

213 "MIT researcher finds evidence of global warming on Neptune's largest moon", News Office, Massachusetts Institute of Technology, June 24, 1998, found at http://web.mit.edu/newsoffice/1998/triton.html.

214 "Pluto is undergoing global warming, researchers find", News Office, Massachusetts Institute of Technology, October 9, 2002, found at http://web.mit.edu/newsoffice/2002/pluto.html.

215 There are also significant El Niño and La Niña effects in the Indian Ocean and, to a lesser extent, in the Atlantic and Southern Oceans.

216 IPCC, *Climate Change 2007*: The Physical Science Basis, Chapter 3: Observations: Surface and Atmospheric Climate Change, Section 3.6.2.1.

217 IPCC, *Climate Change 2007*: The Physical Science Basis, Chapter 8: Climate Models and Their Evaluation, Section 8.4. Among other deficiencies, IPCC climate models fail to predict the timing (except for El Niño) and climatic effects of either the El Niño-Southern Oscillation, which includes both the El Niño warm phase and the La Niña cool phase, or the Pacific Decadal Oscillation, or the Madden-Julian Oscillation.

218 Roy W. Spencer, "Global warming as a natural response to cloud changes associated with the Pacific Decadal Oscillation (PDO)", found at http://www.drroyspencer.com/research-articles/global-warming-as-a-natural-response/.

219 For a summary, see Joe D'Aleo, "US temperatures and climate factors since 1895", at http://icecap.us/images/uploads/US_Temperatures_and_Climate_Factors_since_1895.pdf. Using newly published data from the U.S. Historical Climatology Network (USHCN), D'Aleo in 2008 noted a strong correlation ($R^2 = 0.85$) between the USHCN temperature record from 1900 to 2007 and the combined indices for the Pacific Decadal Oscillation (PDO) and the Atlantic Multidecadal Oscillation (AMO). The PDO and AMO indices are usually based on sea surface temperatures.

220 This was also demonstrated over 30 years ago, says Spencer, by Klaus Hasselman [Claude Frankignoul and Klaus Hasselmann, "Stochastic climate models. Part II: Application to sea-surface temperature anomalies and thermocline variability", *Tellus* 29, pp. 289-305 (1977)]. It comes about because the oceans retain a memory of past changes in the Earth's radiative budget for a very long period, due to their large heat capacity.

221 In this simple energy balance model, the two principal assumptions are the slab ocean depth, through which heat is mixed on multidecadal to centennial time scales, and cloud cover variations directly proportional to the observed PDO index values (see Appendix B: The PDO Cloud Fluctuation Model). A typical ocean mixing depth in the simulations was 800 meters.

222 For calculated values of global warming caused by CO_2, see Table A.1 (Appendix A: Climate Feedbacks and Sensitivity). The 75% and 25% estimates are based on the IPCC's value of 0.8° Celsius (1.4° Fahrenheit) for the increase in global surface temperatures since 1850, rather than the urban heat island corrected value of 0.6° Celsius (1.1° Fahrenheit) discussed in Chapter 3: Science Gone Wrong. If heat island corrected values are used instead of the IPCC numbers for global temperature, Spencer's PDO cloud fluctuation model accounts for almost all of the observed warming.

223 Willie W.-H. Soon, "Variable solar irradiance as a plausible agent for multidecadal variations in the Arctic-wide surface air temperature record of the past 130 years", *Geophysical Research Letters* 32, p. L16712 (2005).

224 See Chapter 5: CO_2 Sense and Sensitivity.

225 IPCC, *Climate Change 2007*: The Physical Science Basis, Chapter 2: Changes in Atmospheric Constituents and in Radiative Forcing, Section 2.7.

226 IPCC, *Climate Change 2007*: The Physical Science Basis, Chapter 9: Understanding and Attributing Climate Change, Section 9.4.1.2.

227 Ibid, Section 9.4.1.4.

Chapter 7: Global Cooling

228 Kyle L. Swanson and Anastasios A. Tsonis, "Has the climate recently shifted?", *Geophysical Research Letters* 36, p. L06711 (2009), available at http://www.uwm.edu/~kswanson/ publications/2008GL037022_all.pdf; reported by Michael Reilly, "Global Warming: On Hold?", *Discovery News*, March 2, 2009, found at http://dsc.discovery.com/ news/2009/03/02/global-warming-pause.html.

229 Ibid. Swanson and Tsonis postulate that global climate undergoes a major shift whenever four specific natural climate cycles resonate in phase, mutually reinforcing one another, at the same time as the coupling between the four cycles increases. The four cycles are the El Niño-Southern Oscillation, the Pacific Decadal Oscillation, the North Atlantic Oscillation and the North Pacific Index. However, the precise cause of this internal reorganization of the climate system is unknown.

230 Satellite global temperature record, which is a weighted mean of land and sea surface temperatures, from the Earth System Science Center at the University of Alabama in Huntsville, reported at http://vortex.nsstc.uah.edu/data/msu/t2lt/uahncdc.lt.

231 HadCRUT3 global combined land and sea surface temperature record, from http://www. cru.uea.ac.uk/cru/info/warming/. The HadCRUT3 temperature record is compiled jointly by the Climatic Research Unit (CRU) and the Met Office Hadley Centre in the UK.

232 If the temperature slide is measured instead from 1998, as some skeptics like to do, the decrease is even larger. But this is only because global temperatures spiked that year, due to the unusually strong El Niño of 1997-1998. When you measure a decline from an abnormally high starting point, you get an artificially big drop. That's data manipulation, of the kind that the alarmist IPCC often indulges in. However, the drop here can legitimately be measured from the beginning of 2002, when the temperature was steadier.

233 Joe D'Aleo, "Correlation of Carbon Dioxide with Temperatures Negative Again", Prison Planet.com blog, May 27, 2008, found at http://www.prisonplanet.com/articles/ may2008/270508_b_carbon.htm.

234 Seth Borenstein, "Obama Left With Little Time to Curb Global Warming", Associated Press, December 14, 2008, found at http://abcnews.go.com/Technology/JustOneThing/WireStory?id=6461910&page=1.

235 Computer climate modelers attribute the cooling that occurred between 1940 and 1970 to aerosol particles in the atmosphere, produced by coal-fired power plants that mushroomed during the period of accelerated industrial development after World War II. According to the models, this cooling effect dominated the warming trend from CO_2 emissions for all those years – although the temperature decline from 1940 to 1970 could also be associated with fluctuations in natural cycles such as the Pacific Decadal Oscillation, something that computer models ignore.

236 Don J. Easterbrook, "Global Cooling is Here: Evidence for Predicting Global Cooling for the Next Three Decades", Global Research, November 2, 2008, found at http://www.globalresearch.ca/index.php?context=va&aid=10783.

237 The cycle time for the Atlantic Multidecadal Oscillation is about 35 years on average, with the warm phases usually lasting longer than the cool phases.

238 See Reference 231 above.

239 See, for example, Theodor Landscheidt, "New Little Ice Age Instead of Global Warming?", at http://bourabai.narod.ru/landscheidt/new-e.htm.

240 IPCC, *Climate Change 2007*: The Physical Science Basis, Chapter 10: Global Climate Projections, Executive Summary.

241 IPCC, *Special Report*: Emissions Scenarios (2000).

242 IPCC, *Climate Change 2007*: The Physical Science Basis, Chapter 10: Global Climate Projections, Section 10.3.1. Over the decade from 2000 to 2010, the average predicted temperature increase for three representative scenarios (Figure 10.5) is approximately 0.2° Celsius (0.36° Fahrenheit); this is equivalent to an increase of 0.14° Celsius (0.25° Fahrenheit) for the seven years from 2002 to early 2009.

243 The IPCC's prediction of a 0.14° Celsius (0.25° Fahrenheit) gain for the period from 2002 to 2009, added to the actual drop of 0.15° Celsius (0.27° Fahrenheit), gives an overestimate of 0.29° Celsius (0.52° Fahrenheit).

244 The temperature increase from 1980 to 2001 was about 0.4° Celsius (0.7° Fahrenheit), according to the IPCC (see Figure 7.2).

245 IPCC, *Climate Change 2007*: The Physical Science Basis, Summary for Policymakers, pp. 13-15.

246 Ibid.

247 IPCC, *Climate Change 2007*: The Physical Science Basis, Chapter 3: Observations: Surface and Atmospheric Climate Change, Executive Summary and FAQ 3.3.

248 Ibid, Table 3.8.

249 Ryan N. Maue, Florida State University, "30-year lows: Global tropical cyclone activity at record low levels", found at http://www.coaps.fsu.edu/~maue/tropical/?p=8. The Accumulated Cyclone Energy (ACE) index is a metric for the collective intensity and duration of tropical cyclones in any particular season. The ACE index measures the sum of the squares of the maximum sustained surface wind speed, at six hourly intervals, for every tropical storm. As seen in Figure 7.3, Maue's analysis shows a recent dramatic drop in the 24-month running sum of the global ACE index.

250 IPCC, *Climate Change 2007*: The Physical Science Basis, Chapter 4: Observations: Changes in Snow, Ice and Frozen Ground, Section 4.4.2.2.

251 Mark Serreze, National Snow and Ice Data Center, reported by *National Geographic News*, September 17, 2008, at http://news.nationalgeographic.com/news/2008/09/080917-sea-ice.html.

252 Some of the sun's incoming radiation is reflected back into space by ice cover, especially near the poles. But once the ice melts, the darker surfaces with lower albedo exposed underneath absorb more solar radiation and contribute to global warming.

253 U.S. National Oceanic and Atmospheric Administration (NOAA), Arctic Report Card 2008, found at http://www.arctic.noaa.gov/reportcard/seaice.html. Summer Arctic ice, which shrinks to its minimum annual extent in September, reached its lowest point in 2007; winter ice, which covers its maximum area in March, was at its lowest level in 2006.

254 See, for example, "Arctic Sea Ice Down to Second-Lowest Extent; Likely Record-Low Volume", Press Room, National Snow and Ice Data Center, October 2, 2008, at http://nsidc.org/news/press/20081002_seaice_pressrelease.html.

255 IPCC, *Climate Change 2007*: The Physical Science Basis, Summary for Policymakers, p. 5.

256 John M. Lyman, Josh K. Willis and Gregory C. Johnson, "Recent cooling of the upper ocean", *Geophysical Research Letters* 33, p. L18604 (2006). The ocean heat content was measured from the surface to a depth of 750 meters.

257 Josh K. Willis, John M. Lyman, Gregory C. Johnson and John Gilson, "Correction to 'Recent cooling of the upper ocean'", *Geophysical Research Letters* 34, p. L16601 (2007).

258 Described in "Argo Robotic Instrument Network Now Covers Most of the Globe", NASA Earth Observatory, December 1, 2004, found at http://earthobservatory.nasa.gov/Newsroom/view.php?id=25723.

259 Josh K. Willis, John M. Lyman, Gregory C. Johnson and John Gilson, "In situ data biases and recent ocean heat content variability", *Journal of Atmospheric and Oceanic Technology* 26, pp. 846–852 (2009).

260 Craig Loehle, "Cooling of the global ocean since 2003", *Energy & Environment* 20, pp. 101–104 (2009). Loehle calculates that the loss of heat from the upper 900 meters of the oceans between 2003 and 2008 was 0.35×10^{22} Joules per year, compared with a gain of 0.92×10^{22} Joules per year by the upper 750 meters from 1993 to 2003, calculated by John Lyman et al (see References 256 and 257 above). From 1955 to 2003, the total gain in heat by the oceans to a depth of 700 meters was 11.2×10^{22} Joules, as estimated by Sydney Levitus et al (in Reference 262 below) – an average gain of 0.23×10^{22} Joules per year.

261 M. Ishii and M. Kimoto, "Reevaluation of historical ocean heat content variations with time-varying XBT and MBT depth bias corrections", *Journal of Oceanography* 65, pp. 287–299 (2009).

262 S. Levitus, J. Antonov and T. Boyer, "Warming of the world ocean, 1955–2003", *Geophysical Research Letters* 32, p. L02604 (2005).

263 Ibid.

264 IPCC, *Climate Change 2007*: The Physical Science Basis, Chapter 5: Observations: Oceanic Climate Change and Sea Level, Executive Summary.

265 Ibid, Section 5.5.6. From 1993 to 2003, thermal expansion of the oceans contributed 60% of the total observed sea level rise.

266 Ibid, Figure 5.13.

267 Ibid, Figure 5.14.

Chapter 8: Why It Matters

268 Tom LoBianco, "Obama climate plan could cost $2 trillion", *The Washington Times*, March 18, 2009, found at http://www.washingtontimes.com/news/2009/mar/18/obama-climate-plan-could-cost-2-trillion/. The estimated cost of the proposed U.S. cap-and-trade system for limiting CO_2 emissions came from congressional staff briefed by the White House, whose original price tag was $646 billion over eight years.

269 The inflation-adjusted cost of World War II to the U.S. has been estimated at $3 - 4 trillion. See, for example, http://www.usatoday.com/news/military/2007-10-23-wacosts_N.htm.

270 United Nations Framework Convention on Climate Change (UNFCCC), Kyoto Protocol (1997), found at http://unfccc.int/kyoto_protocol/items/2830.php. The targeted greenhouse gases are CO_2. methane (CH_4), nitrous oxide (N_2O), sulfur hexafluoride (SF_6), hydrofluorocarbons and perfluorocarbons.

271 One tonne or metric ton is 1,000 kilograms – about 10% larger than a U.S. or short ton.

272 China overtook the U.S. in annual CO_2 emissions in 2008, and is now the world's largest CO_2 emitter. India is not far behind the U.S.

273 U.S. Congressional Budget Office (CBO), "Who Gains and Who Pays Under Carbon Allowance Trading? The Distributional Effects of Alternative Policy Designs", June, 2000, reported in "Containing the Cost of a Cap-and-Trade Program for Carbon Dioxide Emissions", May 20, 2008, available at http://www.cbo.gov/ftpdocs/92xx/doc9276/05-20-Cap_Trade_Testimony.1.1.shtml.

274 Ibid.

275 United Nations Framework Convention on Climate Change (UNFCCC), Convention Text (1992), Article 2, found at http://unfccc.int/resource/docs/convkp/conveng.pdf.

276 At prevailing exchange rates in early 2009, a price of €30 per metric ton was equivalent to around $36 per U.S. ton.

277 Carbon Positive, Carbon trading prices, reported at http://www.carbonpositive.net/searchnewsarticles.aspx?menu=1&s=1&categoryID=3&results=10.

278 Climate Action Network – Europe, "National Allocation Plans 2005-7: Do They Deliver?", Summary for policy-makers, Table 1, available at http://www.climnet.org/EUenergy/ET/NAPsReport_Summary0306.pdf.

279 Julian Glover, "A collapsing carbon market makes mega-pollution cheap", *The Guardian*, February 23, 2009, found at http://www.guardian.co.uk/commentisfree/2009/feb/23/glover-carbon-market-pollution.

280 See, for example, "German inflation hits 3.0 percent in November", AFP, November 27, 2007, at http://afp.google.com/article/ALeqM5jCjNSHDLboaFWO1DMtMKJWtTgiXA.

281 European Union, "Questions and Answers on the Commission's proposal to revise the EU Emissions Trading System", January 23, 2008, found at http://europa.eu/rapid/pressReleasesAction.do?reference=MEMO/08/35&format=HTML&aged=0&language=EN&guiLanguage=en.

282 See "The Lieberman-Warner Climate Security Act (S. 2191)", at http://lieberman.senate.gov/documents/lwcsaonepage.pdf.

283 U.S. Environmental Protection Agency (EPA), "Acid Rain and Related Programs: 2007 Progress Report", found at http://www.epa.gov/airmarkt/progress/docs/2007ARPReport.pdf.

284 U.S. Environmental Protection Agency (EPA), "Analysis of the Lieberman-Warner Climate Security Act of 2008", March 14, 2008, found at http://www.epa.gov/climatechange/downloads/s2191_EPA_Analysis.pdf.

285 Ibid.

286 U.S. National Association of Manufacturers (NAM), "United States: Economic Impact from the Lieberman-Warner Proposed Legislation to Reduce Greenhouse Gas Emissions", available at http://www.accf.org/pdf/NAM/ACCF-NAM-US.pdf.

287 The NAM defines low-income households as those with average annual incomes less than $18,500.

288 See, for example, CRA International, Inc., "Economic Analysis of the Lieberman-Warner Climate Security Act of 2007 Using CRA's MRN-NEEM Model", at http://www.crai.com/ uploadedFiles/RELATING_MATERIALS/Publications/BC/Energy_and_Environment/ files/CRA_NMA_S2191_April08_2008.pdf. CRA International is an independent business consulting firm.

289 Grant Fleming, "Emissions trading scheme up for review under Act deal", *The New Zealand Herald*, November 16, 2008, found at http://www.nzherald.co.nz/politics/news/ article.cfm?c_id=280&objectid=10543330.

290 National-ACT Confidence and Supply Agreement, November 16, 2008, found at http:// www.national.org.nz/files/agreements/National-Act_Agreement.pdf.

291 Ibid.

292 James Grubel, "Australia parliament debates carbon-trade laws", Reuters UK, May 14, 2009, at http://uk.reuters.com/article/oilRpt/idUKSYD46220220090514?pageNumber=1 &virtualBrandChannel=0.

293 "Carbon tax", at http://en.wikipedia.org/wiki/Carbon_tax.

294 "CO_2 emissions booming, shifting east, researchers report", Oak Ridge National Laboratory News Release, September 24, 2008, found at http://www.ornl.gov/info/press_ releases/get_press_release.cfm?ReleaseNumber=mr20080924-00.

295 U.S. Department of Energy (DOE), "20% Wind Energy by 2030: Increasing Wind Energy's Contribution to U.S. Electricity Supply", July 2008, found at http://www1.eere.energy.gov/ windandhydro/pdfs/41869.pdf.

296 Based on statistics compiled by the U.S. Department of Energy (DOE)'s Energy Information Administration, reported at http://www.eia.doe.gov/international/ RecentTotalElectricConsumption.xls.

297 According to the DOE study (in Reference 295 above), generating 20% of U.S. electricity from wind by 2030 could avoid approximately 825 million metric tons (750 million U.S. tons) of CO_2 emissions from the electric sector, out of a projected 7,900 million metric tons (7,200 million U.S. tons) in total U.S. CO_2 emissions. The Energy Information Administration (in Reference 298 below) estimates total CO_2 emissions in 2030 to be 6,400 million metric tons (5,800 million U.S. tons).

298 U.S. Department of Energy (DOE), Energy Information Administration, "Annual Energy Outlook 2009", March 2009, p. 84, found at http://www.eia.doe.gov/oiaf/aeo/pdf/0383(2009).pdf.

299 Michael J. Trebilcock, "Wind power is a complete disaster", *The National Post*, April 8, 2008, found at http://network.nationalpost.com/np/blogs/fpcomment/archive/2009/04/08/wind-power-is-a-complete-disaster.aspx.

300 "Danish energy consumption and emissions increased in 2006, report finds", *Energy Business Review*, April 23, 2007, found at http://www.energy-business-review.com/article_news.asp?guid=47017FA7-BEF9-4E82-B994-F36188328983. Following the 16% rise in 2006, Danish CO_2 emissions fell in 2007 – but emissions that year were still above their 2005 level.

301 See Reference 298 above, p. 74.

302 Ibid, p. 71. Coal is expected to contribute 47% of U.S. electricity demand in 2030, and natural gas 20%.

303 Gabriel Calzada Álvarez, "Study of the effects on employment of public aid to renewable energy sources", March 2009, available at http://www.juandemariana.org/pdf/090327-employment-public-aid-renewable.pdf.

304 U.S. Department of Energy (DOE), Energy Information Administration, "Federal Financial Interventions and Subsidies in Energy Markets 2007", April 2008, Chapter 5, Table 35, found at http://www.eia.doe.gov/oiaf/servicerpt/subsidy2/pdf/chap5.pdf.

Chapter 9: Summary

305 This joke was originally part of a letter from Bobby Henderson to the Kansas School Board in the U.S., illustrating the point that correlation does not imply causation. The 2005 letter, which is a spoof of intelligent design theory, can be found at http://www.venganza.org/about/open-letter/.

Appendix A: Climate Feedbacks and Sensitivity

306 IPCC, *Climate Change 2007*: The Physical Science Basis, Annex I (Glossary).

307 IPCC, *Climate Change 2007*: The Physical Science Basis, Chapter 8: Climate Models and Their Evaluation, Section 8.6.2.3.

308 IPCC, *Climate Change 2001*: The Scientific Basis, Chapter 6: Radiative Forcing of Climate Change, Equation (6.1).

309 Ibid, Table 6.2.

310 IPCC, *Climate Change 2007*: The Physical Science Basis, Summary for Policymakers, p. 4.

311 IPCC, *Climate Change 2001*: The Scientific Basis, Chapter 9: Projections of Future Climate Change, Table 9.A1.

312 IPCC, Ibid, Executive Summary.

313 IPCC, *Climate Change 2007*: The Physical Science Basis, Chapter 9: Understanding and Attributing Climate Change, Section 9.6.2.1.

314 Ibid, Section 9.4.1.2.

315 IPCC, *Climate Change 2007*: The Physical Science Basis, Chapter 10: Global Climate Projections, Box 10.2.

316 James Hansen et al, "Earth's energy imbalance: confirmation and implications", *Science* 308, pp. 1431-1435 (2005), and "Climate simulations for 1880–2003 with GISS model E", *Climate Dynamics* 29, pp. 661–696 (2007).

317 Roy W. Spencer, "Global warming as a natural response to cloud changes associated with the Pacific Decadal Oscillation (PDO)", found at http://www.drroyspencer.com/research-articles/global-warming-as-a-natural-response/. The calculated climate sensitivities of 0.21°C for today and 0.45°C for doubled CO_2 correspond to the maximum feedback parameter of 8.3 W/m^2 per °C.

318 The zero feedback (Planck) feedback parameter calculated from computer climate models has a value of 3.2 W/m^2 per °C, according to Soden and Held [Brian J. Soden and Isaac M. Held, "An assessment of climate feedbacks in coupled ocean–atmosphere models", *Journal of Climate* 19, pp. 3354-3360 (2006)]. A slightly different value of 3.3 W/m^2 per °C is reported by Forster and Taylor [Piers M. de F. Forster and Karl E. Taylor, "Climate forcings and climate sensitivities diagnosed from coupled climate model integrations", *Journal of Climate* 19, pp. 6181-6191 (2006)].

319 IPCC, *Climate Change 2007*: The Physical Science Basis, Summary for Policymakers, p. 2.

320 IPCC, *Climate Change 2001*: The Scientific Basis, Chapter 8: Model Evaluation, Table 8.1.

321 IPCC, *Climate Change 2007*: The Physical Science Basis, Chapter 8: Climate Models and Their Evaluation, Table 8.1.

322 IPCC, *Climate Change 2007*: The Physical Science Basis, Chapter 9: Understanding and Attributing Climate Change, Section 9.6.3.2.

323 IPCC, *Climate Change 2007*: The Physical Science Basis, Summary for Policymakers, p. 5.

324 Sandrine Bony et al, "How well do we understand and evaluate climate change feedback processes?", *Journal of Climate* 19, pp. 3445-3482 (2006), found at http://www.atmos.ucla.edu/csrl/publications/Hall/Bony_et_al_2006.pdf.

325 R. A. Colman, S. B. Power and B. J. McAvaney, "Non-linear climate feedback analysis in an atmospheric general circulation model", *Climate Dynamics* 13, pp. 717–696 (1997).

Appendix B: The PDO Cloud Fluctuation Model

326 Roy W. Spencer, "Global warming as a natural response to cloud changes associated with the Pacific Decadal Oscillation (PDO)", found at http://www.drroyspencer.com/research-articles/global-warming-as-a-natural-response/.

327 The PDO index is a measure of Pacific sea surface temperatures.

328 HadCRUT3 temperature record, available at http://www.cru.uea.ac.uk/cru/data/temperature/.

329 Outgoing longwave (LW) and reflected shortwave (SW) radiative fluxes were measured by the CERES instrument on NASA's Terra satellite.

330 Roy W. Spencer and William D. Braswell, "Potential biases in feedback diagnosis from observational data: A simple model demonstration", *Journal of Climate* 21, pp. 5624-5628 (2008).

Index

consensus, lack of scientific, 3-5, 8, 122

corruption, IPCC, 6, 20, 24-31, 36-45, 74-77, 122

cosmic rays, 12, 55, 74, 80-81, 85, 88, 120-121

cosmogenic isotopes, 75-76, 80

cost, of CO_2 regulation, 101-116, 122-123; of electricity, 108, 116; of renewable energy sources, 110, 113-116; wasted money, 103, 107, 109, 111

D'Aleo, Joe, 86, 92

Dansgaard-Oeschger events, 77-78

data contamination, of temperatures by human activity, 21-23, 35, 36-40

data fabrication, 20, 22, 36-40, 42, 120, 122

data manipulation, IX, 19-34, 67, 77, 119-122

deception, by IPCC, 1, 24-31, 34-35, 42, 67, 76-77, 88, 119, 122

Deming, David, 30

Denmark, 112-113, 123

diminution, *see* feedback

DOE, 39-40, 112, 114-116, 135

doubled CO_2, as measure of climate sensitivity, 18, 54, 61, 64-65, 118, 126-129

Earth's orbit, 11, 68, 73

East Antarctica, 57

economic analysis, of temperature data, 21-22, 37

economic consequences, of CO_2 regulation, 103-111, 113, 122-123; for

jobs, 104, 108, 113, 115-116, 123; on manufacturing, 108, 115; of switching to renewables, 113, 115-116, 123

El Niño, 11, 50, 52, 55, 84, 135

electricity, contribution from fossil fuels in U.S., 111-112, 114, 123; contribution from wind energy, 112-116, 123; utility companies, 105, 108-109, 112, 123

emissions trading, *see* cap-and-trade

energy prices, 104, 108, 113, 116, 123

environmentalism, VII, 1, 7, 103, 110, 112, 115, 123

EPA, 108-109, 135

European Union (EU), 105-107, 112-115, 123; failure of EU cap-and-trade, 105-107, 123

feedback, 60-63, 125-130, 135; amplification (positive feedback), 16, 18, 60-65, 68-72, 118, 127-130; diminution (negative feedback), 16, 60-66, 87, 118, 127-130, 132; parameter, 127-130, 132; net, 60-61, 63, 65, 71-72; nonlinear, 130

forcing, climate, 37, 64, 125-126, 132-133, 136

fossil fuels, as backup to renewables, 112-114, 123; burning of, 1, 5, 13, 57, 103, 105, 107, 110-114, 117, 136

Fourth Assessment Report, IPCC, 7, 22, 30, 37, 40, 45, 54, 70, 75-76, 95, 97, 99, 125

fraud, 36, 38-40, 120, 122

glaciers, 11, 68, 78, 97-98, 100